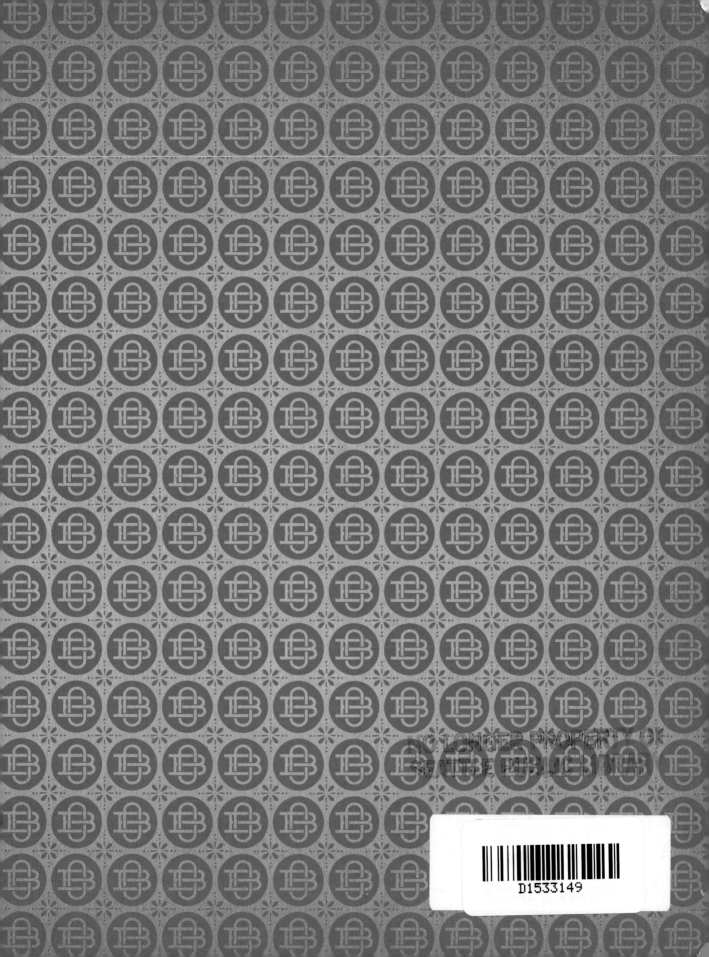

GAIL ANDERSON

FOREWORD BY DEBBIE MILLMAN

tside the Box

the Box

HAND-DRAWN PACKAGING FROM AROUND THE WORLD

PRINCETON
ARCHITECTURAL
PRESS
NEW YORK

For my favorite seniors, Lloyd and Nola Anderson

PUBLISHED BY
Princeton Architectural Press
37 East Seventh Street
New York, New York 10003

Visit our website at www.papress.com.

EDITOR: Tom Cho
DESIGNER: Anderson Newton Design

Special thanks to: Nicola Bednarek Brower, Janet Behning, Erin Cain, Megan Carey, Carina Cha,
Andrea Chlad, Tom Cho, Barbara Darko, Benjamin English, Jan Cigliano Hartman, Jan Haux,
Lia Hunt, Mia Johnson, Stephanie, Leke, Diane Levinson, Jennifer Lippert, Jaime Nelson,
Rob Shaeffer, Marielle Suba, Kaymar Thomas, Paul Wagner, Joseph Weston, and Janet Wong of
Princeton Architectural Press

—Kevin C. Lippert, PUBLISHER

Library of Congress Cataloging-in-Publication Data
Anderson, Gail, 1962–
Outside the box : hand-drawn packaging from around the world / Gail Anderson. — First edition.
    pages cm
ISBN 978-1-61689-336-1 (paperback)
1. Packaging — Design — Themes, motives. I. Title.
NC1002.P33A53 2015
741.6—dc23

Contents

LETTERS OF LOVE
Debbie Millman

The first love letter I ever wrote was to a boy who sat behind me in my tenth grade trigonometry class. I was terrified to reveal my true feelings, so rather than pass a note to him in class or ask a friend to tell him, I decided to send an anonymous message. Late one afternoon, when no one was looking, I went up to his locker, took a pencil out of my handbag, and wrote right on the locker door: *I like you.* And then, for added drama and flourish, I drew a heart around my impulsive communiqué. My silly high school high jinks caused quite a stir among our friends, including the boy, who spent hours trying to figure out who could have left the mysterious message. My opinion on the matter was sought out, which was rather unfortunate, since no one considered me the possible culprit. Needless to say, I never revealed my authorship to anyone.

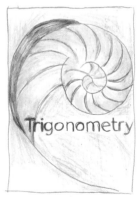

Since then, most of my sentimental gestures have been more forthright, and any pronouncements sent in return have been meticulously treasured. I have kept many of the love letters of my youth safely tucked away in fabric boxes. I've also kept an assorted collection of other tidings, including my sixth grade autograph book and hastily written postcards from friends in foreign countries. I even managed to keep the first handwritten phone message, taken by the receptionist at a job I had in 1987, from the man who eventually became my first husband. Sadly, the message lasted longer than the marriage.

Whether through letters, cave drawings, or handwritten phone messages, I believe that humans have an innate and nearly involuntary need to communicate with one another. We are compelled to share stories, reveal secrets, compare notes, and mark time.

The Swiss linguist Ferdinand de Saussure stated that spoken language is an expression of thought, and that written language is purely a supplement to vocalization. He believed that writing was to be regarded secondary to speech, and that speech alone held the center of human attention. But as technology has become the centerpiece of our daily lives, and with the increased dominance of the media, signs and symbols have begun to take the place that spoken language once long held in our culture. Linguists believe that speech is due to a semiotic shift, and that spoken language will no longer occupy the seat of privilege, as writing moves into the spotlight, with its sign-based

symbolism. So while our prehistoric ancestors lived only in a physical universe, we now inhabit a more symbolic realm. Language, myth, art, and, religion are now all part of this experience and have become threads woven together into a net of our shared human experience.

I can't help but wonder what the cognitive effect of this mode of communication is. Our love affair with communicating by keyboard has drastically reduced the amount we write by hand, so much so that the *New York Times* has reported that the skill, "like an unused muscle," is pretty much dead by the time we reach high school. *High school!*

What are the ramifications of losing our handmade muscle? Our current ability for prolific communication is certainly collectively beneficial to the culture, but what does it mean for us individually? Throughout time, we have used our hands to satisfy our needs—whether spiritual or practical. The creation of meaning from nothing may be our greatest achievement. It bears witness to the artist—and the human—in all of us. And while computers might set type in near flawlessly accurate columns, what we make by hand is beautiful by virtue of its irregularity and integrity and soul.

Nowhere is this more evident than in the stunning pages of Gail Anderson's new book, *Outside the Box: Hand-Drawn Packaging from Around the World.* The hand-drawn typography and packaging created by designers such as Michael Bierut, Dana Tanamachi-Williams, and Jon Contino combine Old and New World aesthetics to create sensuous, shelf-stopping product packaging far more exciting and appetite appealing than any typeset label ever possibly could.

What resonates in these packages is an inherent authenticity and honesty. Like a fingerprint, the visual language of these hand-drawn labels provides an indelible imprint. What is contained in this book is both beautiful and enduring. Yes, these packages are printed, but that is the allure of the hand-drawn package: it is a magical combination that reconciles the duality of printed word and painted face, creating something that is real and well-intended. For the first time ever, Outside the Box focuses on the makers behind the typography and the package design. The artists featured here are no longer anonymous; their work is very much beloved. The results present products with an enduring, uniquely human spirit. As we live our lives and search for meaning (and love), it is these handcrafted messages that have the magnitude—and the soul—to measure, reflect, and express who we are, and fulfill our deepest desires.

Justo Botanica's handmade and printed labels date back in some cases to the 1940s, when owner Jorge Vargas's stepfather ran the shop.

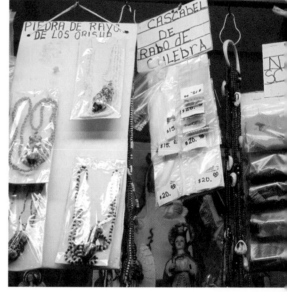

INTRODUCTION

My infatuation with hand-drawn type began about five years ago while visiting Justo Botanica, in New York City's El Barrio in East Harlem. I'd ride my bike over to visit with Jorge Vargas, the shop's proprietor and neighborhood herbalist, and before long, I'd find myself photographing or buying many of his fragrant remedies, prayer candles, and oils.

Fast forward to the present and I could probably open my own shop with the amount of stuff I've picked up at Justo. My intention was not to ward off evil spirits, though that certainly couldn't hurt. I was enamored of the hand-drawn type that Jorge and his stepfather before him created for their packages. On a random visit, he asked what I thought of the idea of using digitally designed labels for his products, as a local graphic designer suggested that he do. I was aghast. Jorge was amused by my reaction— and probably a bit confused, since, after all, wasn't I one of those graphic designers with a fancy computer?

Justo Botanica's earnest, often crudely handwritten and illustrated packages are right up my alley. I love the completely unselfconscious way in which they were created, and are then marketed in the shop. Jorge uses a two-color, hand-cranked printer—I couldn't tell you what it's called— and photocopies to create his artwork. And while he laughs off my horror at the thought of "going digital," I think he takes great pride in the heritage and authenticity of the work he's created by hand.

Another collection of hand-drawn packaging that I photographed, and

then bought (of course), was from Jamie Oliver's Jme line of home products. While its creator, Pearlfisher, had no sketches available for this project, I felt the book would be incomplete without its inclusion. Pearlfisher was tasked with creating a new brand and lifestyle concept for celebrity chef Jamie Oliver, which would transport the Jamie Oliver brand from the restaurant to the home. The strikingly original packaging for this eclectic lifestyle brand reflects the specific function of each product, while the bold Jme logo holds the collection together. As Pearlfisher describes it, "Using the idea of taking the Jamie Oliver experience from the kitchen to the home, the hand-drawn design was inspired by… hand-drawn labels on the jars of homemade jam and the fresh produce found at farmers markets." Tasty.

"From passing trend to established métier, hand-inscribed packages draw their inspiration from the kitchen table jams, jellies, and liquors that were once appealingly quaint and now are professionally appealing," says author and School of Visual Arts MFA Design cochair Steven Heller (though he fails to mention Justo Botanica, so I'll have to take him there some day). The growing contemporary movement has extended beyond stationery and posters all the way to advertising and, of course, packaging.

"The genius of hand lettering is that the final product must look spontaneous, emotional, and energetic, but the development of it must be meticulous, logical, and controlled," says Maryann Mitkowski, vice president and director

"In all my years of seeing packaging trends come and go, there is one style that has stood the ultimate test of time: hand-drawn."

— ANDREW GIBBS, founder and CEO, The Dieline

PRODUCT: Jme
CLIENT: Jamie Oliver
DESIGN FIRM: Pearlfisher
ART DIRECTOR/HAND LETTERER: Natalie Chung
DESIGNER: Sarah Pidgeon
MEDIUM: Pencil
COUNTRY: United Kingdom

of Creative Services for Parham Santana, the New York City–based brand extension agency. "Every counter space, stress, and stroke telegraphs a specific positioning or attitude. It takes a masterful hand to get it right."

Mitkowski continues: "Now more than ever, we need to balance our constant barrage of digitized images with the unique imperfections that come from the human hand. Of all the beautiful typefaces in the world, none puts the heart on the page as quickly as hand-drawn type."

Andrew Gibbs, founder and CEO of The Dieline, the world's largest packaging blog, adds, "In all my years of seeing packaging trends come and go, there is one style that has stood the ultimate test of time: hand-drawn. In the past few years, I have seen this style become even more popular on a global scale. After the recession, people's shopping habits changed. They bought less consumer products, especially products that were considered excessive or unnecessary. Consumers began to demand products that were authentic, honest, and simple. Brands had to respond to this quick consumer shift, and many responded by returning to the authentic core of their brand."

"For a lot of companies," Gibbs continues, "their authenticity was best visually represented by hand-drawn elements. This is because there is nothing that can replicate the artistry of handmade lettering and hand-illustrated graphics. Hand-drawn designs are the ultimate expression of authenticity and humanity, which is why it is so effective across so many products, and so effective over the course of time."

Are there typefaces that can replicate

hand-drawn type? Certainly—some implemented so convincingly, it made weeding through projects for this book more difficult than anticipated. It's tempting to set secondary and legal type in an existing hand-drawn font, but *Outside the Box* is filled instead with designers crazy enough to draw their own barcodes, as well as all the tertiary type.

"Hand-drawn type tells you instantly that the product is personal and unique," says New York City–based designer Roberto de Vicq de Cumptich. "No two hands could have created the same graphics, the same mistakes, the same inconsistencies—and so the product is distinctive in every way. It gives it an artisanal quality and character in our increasingly mechanized world where everyone has access to fonts and computers, which are relatively cheap and abundant. Hey, even McDonald's is using hand-drawn type! The trick is to make it relevant, appropriate, clever, and so seductively gorgeous."

The designers within these pages employ a broad range of skills, from crude and naïve to masterful execution of traditional techniques. Despite each design being carefully crafted to target a specific audience, I think you'll find all of them compelling in their own unique way. Much to my delight, the designers have generously pulled back the curtain to share their process sketches, allowing us to appreciate their approach to each design challenge as well as their final results. I hope the work will inspire you to whip out your own pencil, Sharpie, paintbrush, or whatever, and keep the hand-drawn love going. —GA

CHILLI SEA SALT

a pinch of ROSE MARY sea salt

she sells sea salt on the sea shore. I'm sure it's Citrus sea salt for sure

LEMON CURD .Jme

seville orange marmalade .Jme

"It wasn't a case of trying out different designers' styles,"
Pearlfisher says. "Rather, it became clear very quickly that the
hand-lettering style worked for the design, and that the style of
the designer whose concept it was suited it very well."

Wikipedia defines "DIY" as "the method of building, modifying, or repairing something without the aid of experts or professionals." I think they got it wrong. DIY is all about empowerment and craft. Can I hand letter that package copy, even though, you know, technically I'm not a hand letterer? Why, yes I can!

The do-it-yourself spirit is alive and well in this section, where old-school art supplies reign supreme.

NUTS.COM
Pentagram

Pentagram is the world's largest independent design consultancy. The firm is owned and run by nineteen partners, a group of friends who are all leaders in their individual creative fields.

Pentagram's offices are located in London, New York, San Francisco, Berlin, and Austin. They design everything: architecture, interiors, products, identities, publications, posters, books, exhibitions, websites, and digital installations.

Every client of theirs works directly with one or more of the firm's partners; Pentagram's conviction is that great design cannot happen without passion, intelligence, and personal commitment, which is reflected in a portfolio of work that spans five decades.

PRODUCT: Nuts.com

CLIENT: Nuts.com

DESIGN FIRM: Pentagram

CREATIVE DIRECTOR:
Michael Bierut

DESIGNERS: Katie Barcelona and Aron Fay

HAND LETTERER: Font design by Jeremy Mickel, based on a hand-drawn alphabet by Michael Bierut

ILLUSTRATOR:
Christoph Niemann

MEDIUM: Ink and digital

COUNTRY: United States

PROJECT DESCRIPTION: Nuts.com is exactly what it sounds like: an online retailer of every kind of nut—in salted, unsalted, and organic varieties—from peanuts, pistachios, pecans, and pine nuts to cashews, almonds, and filberts. True to their slogan, "We're more than just nuts," Nuts.com also offers dried fruit, snacks, chocolate, coffee, and tea.

The company first launched its site in 1999 with the slightly confusing name/URL NutsOnline.com—their second choice after Nuts.com, which had already been taken at the time. In 2014, the company finally secured the Nuts.com address and made the move to the new name. With the change, the company asked Pentagram's Michael Bierut to create a new identity and packaging that would establish Nuts.com as a distinctive brand. The new graphics create an unmistakable look and feel that is fun, personal, and a little nutty. Bierut's redesign incorporates this irreverent and chatty tone of voice into the packaging with friendly hand-drawn typography and illustrations. The result is playful, inviting, and instantly memorable.

How much of this was set using the typeface (which you originally painted by hand) and how much is custom?
MICHAEL BIERUT: All of it was done using the typeface. Once we had it digitized, the game was to see how crazy we could get without picking the paintbrush back up. There were a lot of alternate characters.

Were other lettering/typeset solutions considered?
Yes, we were originally looking at multiple directions. My favorite of the alternates was a customized Cooper Black with all the letters colored and modeled to look like nuts. It made sense as a concept, I suppose, but the overall effect just looked like sugary breakfast cereal for kids.

Did font designer Jeremy Mickel work from the type sample (shown) or did he create all of the alternate characters?

All of the characters, including the alternates, are based on letters that I hand painted.

How easy (or difficult) was it to get the client's buy-in?
With the new URL secured, fledgling Nuts.com CEO Jeff Braverman and his team were really ready for something different. They also have a very secure sense of who they are, and are very confident about their tone of voice and their relationship with their customers. And best of all, because the product is sold exclusively online, the packaging isn't faced with the challenge of competing on the grocery store shelf. Their customers buy nuts based on the pictures of nuts. The box and the product packaging are more like gift wrapping, which is how Jeff thinks of their fulfillment process: sending the customers wonderful gifts.

OPPOSITE: Redesigned Nuts.com packaging featuring the bold new identity

The original NutsOnline.com packaging is paired with
its updated counterpart. The difference is striking.

The Nuts.com typeface is based on an alphabet hand drawn by Bierut.

AAαABBBBBCCCCCDDDαD
EEEeeFFFFFGGGgggHHHHh
IIIIIJJJJJKKKKKLLLLL
MMMMMNNNnOOOOOPPPP
QQQQQRRRRRSSSSSTTTTt
UUUUUVVVVVWWWWW
XXXXXyyyyYYzzzzZ

Nuts Nuts Nuts Nuts
Nuts Nuts Nuts Nuts
Nuts Nuts Nuts Nuts

Nuts.com   Nuts.com

getting seriously nutty since 1929

We're Nuts

It's Nuts in Here

TOPSHOP MAKE UP
Sarah Thorne

Sarah Thorne is an independent UK-based designer working on logotypes, identities, packaging, print, publishing, point-of-sale materials, products, and websites for individuals and larger brands. Thorne's design experience crosses fashion, lifestyle, and luxury sectors with a love of creative formats, diverse print processes, and unique materials.

PRODUCT: Topshop Make Up
CLIENT: Topshop
DESIGN FIRM: Sarahthorne.co.uk
ART DIRECTOR/DESIGNER/HAND
LETTERER/ILLUSTRATOR:
Sarah Thorne
PHOTOGRAPHER: Adam Laycock
MEDIUM: Crayon
COUNTRY: United Kingdom

PROJECT DESCRIPTION: Thorne designed the packaging for Topshop's makeup range, using black boxes with illustrations of the products on the front, hand-lettered titles, and cosmetics decorated with patterns.

You managed to capture a lot of the graphic look of pencil/chalk. Are there any particular production concerns in terms of creating lines that are both soft/organic and crisp and graphic?

SARAH THORNE: I spent a long time getting that "chalky" look; in fact I mostly used a Caran d'Ache crayon. I was determined not to lose the quality of the crayon, as I knew this would go a long way in making the packaging convincing. My artwork was actually "corrected" by our well-meaning printer, who took away what they thought were the imperfections. When all the proofs came back "smoothed out," it was a challenge to explain that we wanted to preserve the original look.

Did you try to evoke some of the energy and layout of your sketchbook pages in the design collateral?

In a way, because I realized while working in my sketchbook that a hand-done approach would be perfect, since makeup is applied by hand, etc. I wanted the packaging to appear as though the makeup itself had been used. I actually set out using the lipsticks, but they were too soft and melted quickly

in my hands, so I used the crayon, which looked very similar in texture, but could be sharpened and was much less greasy.

Did you hand letter all the small print as well? Or did you composite or create a font?

I hand lettered everything at first, but then, as so much needed that "language," I went on to digitize each letter. It all had to be scanned at very high resolution and painstakingly traced in Illustrator so as to not lose any of the "noise." The very small type on the packaging was Gotham; it was too risky to hand letter the ingredients lists.

How has the campaign evolved from season to season?

The branding has essentially remained the same, with capsule "trend" ranges appearing seasonally. For these seasonal ranges, Topshop provides me with mood boards and a theme to work to, but I maintain the continuity of the main line and still use that same crayon when decorating new packaging.

The stark black-and-white elegance of the cosmetic line is turned upside down by contrasting it with Thorne's cheerful lines and dots.

LAUNCHES

TOPSHOP MAKE UP

MAKE THE 5TH WAY

YES! WE'VE EXACTLY WHAT CREATED MAKE WAITING FOR. WHAT YOUR MAKE MAIN LINE. IN MAKE UP UP ESSENTIAL COLLECTION, ADDITION, BAG AND IT'S

Topshop's affordable cosmetics are designed to appeal to a youthful audience.

Thorne achieved her "chalky" look using a Caran d'Ache crayon.

DÎNER ST. VALENTIN

PANTONE® 426 U	PANTONE® Black 7 M	PANTONE® 426 M

LES PETITES

TOPSHOP

Dos de cabillou

Aioli et ses légu

Moelleux aux

EYES
LINER MASCARA

TOP TEN
MOMENTS IN THE HISTORY OF

MASCARA

ITS MORE FUN
TO BE A PAINTER
THAN THE paint
IF GEORGE
CLOONEY CAN'T
SAVE THE WORLD AT least
HE MAY BE ABLE TO SAVE
THE MOVIES
PHOTOGRAPH BY CHRIS JONES

PENCIL

colour ref - 426u/404u
Pantone too grey - need more half tone?

EYES

OFF WHITE ON BLACK
EXTERNAL PKG.

BLACK ON BEIGE.
INTERNAL PKG.

brighten crayon

VALTIDA PICCOLA
Bruketa&Žinić OM

Bruketa&Žinić OM is a group of marketing communication agencies with offices in Vienna, Austria; Zagreb, Croatia; Belgrade, Serbia; and Baku, Azerbaijan. Some of the agencies included are Brandoctor brand consultancy, Brlog digital agency, and Brigada—a studio that designs retail, exhibition, and office spaces, as well as products and architecture. The group has been named International Small Agency of the Year by *Advertising Age* in 2013, and awarded the second place prize as the Most Efficient Independent Agency, Globally, by Effie Worldwide in 2012.

PRODUCT: Valtida Piccola
CLIENT: Valtida Piccola
DESIGN FIRM: Bruketa&Žinić OM
CREATIVE DIRECTORS:
Davor Bruketa and Nikola Žinić
ART DIRECTOR/DESIGNER/HAND
LETTERER: Nebojsa Cvetkovic
PHOTOGRAPHER: Domagoj Kunić
PRODUCTION MANAGER:
Vesna Durasin
PRE-PRESS: Radovan Radicevic
MEDIUM: Watercolor
COUNTRY: Croatia

PROJECT DESCRIPTION: Valtida Piccola is a brand of olive oil from Istria, Croatia. The goal was to differentiate the product from similar ones on the market, to create something different from typical olive oil packaging, accentuating the fact that it is a handmade product.

There is an exuberance to this packaging. Can you talk about the playful direction you chose?
NEBOJSA CVETKOVIC: The idea was derived from the fact that the oil is handmade. We combined large, hand-drawn lettering with fine, subtle serif typography to convey that distinctive duality of the handmade aspect of the product and professional quality of the brand.

How does the hand-drawn trend play itself out in Croatia?
The hand-drawn lettering trend is very popular in Croatia, and has found its way into mainstream commercial design and advertising.

Why did you choose rough edges for the type rather than smooth?
We just tried to make it as sketchy and raw as possible, to accentuate the "organic" feel of the product.

How did you choose the accompanying serif typeface that you used?
You can never go wrong with Garamond.

OPPOSITE: *Bruketa&Žinić succeeded in differentiating Valtida Piccola from other olive oils by using hand-drawn type to reinforce its status as a handmade product.*

Valtida Piccola
*Ekstra djevičansko
maslinovo ulje*

VALTIDA
PICCOLA
OLIO
EXTRA
VERGINE
D'OLIVA
½L

Berba 2012.
Rovinj, Hrvatska

Each watercolor sketch is energetic yet tasteful in its simplicity. The overall feel is contemporary and not typical of olive oil packaging on store shelves.

VALTIDA
PICCOLA
olio extra
vergine
d'oliva

- Valtida -
piccola
OLIO
EXTRA
VERGINE
D'OLIVA

VALTIDA
PICCOLA
OLIO
EXTRA
VERGINE
D'OLIVA

Valtida Piccola

Olio extra
vergine d'oliva
Extra djevičansko
maslinovo ulje

½ l

VALTIDA
PICCOLA
EXTRA
VERGINE
OLIO
D'OLIVA
½ L

33

FOODIE GARDEN
Joel Holland

J oel Holland is a hand letterer and illustrator working with individuals and companies from around the world. Clients include Apple, Wieden+Kennedy, Random House, Little, Brown & Co., *New York* magazine, and the *New Yorker*. His work has been recognized by American Illustration, the AIGA, and the Art Director's Club, and has appeared on book jackets, billboards, and even a model's butt.

PRODUCT: Foodie Garden

CLIENT: Noted

DESIGN FIRM: Joel Holland Illustration

ART DIRECTOR: Doug Woiske

HAND LETTERER/ILLUSTRATOR: Joel Holland

MEDIUM: Ink, paper, and silk screen

COUNTRY: United States

PROJECT DESCRIPTION: Foodie Garden is a collection of seed-growing kits for home cooks looking to push the flavor envelope. All that you need is included: the pot, the peat, and the seeds. It's a way to get more bang for your flavor buck right from your windowsill!

Your lettering covers a pretty broad range of type styles, from ornate scripts and ornamented wood to simple sans. How do you choose the right style for each project?

JOEL HOLLAND: I let the context determine the style. In this instance, they wanted an approach that was louder and more alive than the competitive products already on the market. My aim was to satisfy that goal while keeping a consistent look.

The color sketch has the background color and slip shadow. It does not appear to have made the final cut. Why not?

I believe it was a choice to go in a more immediate direction: less ink, less labored. I was psyched about it because I'm a big fan of the one-color produce boxes you see on the streets of New York. This mirrors that aesthetic.

Can you walk through your lettering process? Do you work by tracing letterforms, or is your work done freehand?

It's all done freehand. I'll comb through a lot of reference material. Books, photos, blogs, etc. This way there's always a freshness and a good variety of styles.

How do you address issues of scale/weight/ legibility between small and large letters?

Again, I go by the context, much like a typesetter would, I suppose. With this project being a series, I wanted to keep the body copy consistent and ultralegible. The display/illustrated areas were naturally going to run larger, so there is a bit more wiggle room with the look of the letters/words.

FOODIE GARDEN
HOT BONNY!
SCOTCH BONNET PEPPERS

Scotch Bonnet Peppers

Mainly founded & used in the Caribbean islands, these BLiSTERiNG balls of FiRE ripen to BLAZiNG orange colour. Be careful not to get burned!

Holland's casually crafted typography helps reinforce Foodie Garden's true DIY spirit.

"It was a pretty quick process from sketch to final," Holland says. "The client was very flexible and clear in their brief. My intention was for the work to feel as raw and real as possible for the context."

FOODIE GARDEN

Basil O Holic

BASILS ARE PERHAPS THE *MOST POPULAR,* *Versatile & Tasty* HERB for COOKING. IN FACT LET'S JUST SAY A *Fresh* SUPPLY of BASIL A COOK DOTH MAKE.

SWEET BASIL

The ITALIAN BASIL. BRIGHT GREEN LEAVES, (Bushy) PLANTS. NEVER MET A TOMATO IT DID NOT L♥VE. ALSO, THE BASIL for *Easy to grow.* EASY TO LOVE.

SIAM QUEEN

SHE'S AN *Exotic* BASIL WITH A HINT OF ANISE. **LARGER LEAVES** & PRETTY PURPLE *Flowers* MAKE THE QUEEN A GREAT *Ornamental* PLANT AS WELL AS A *delightful herb* TO COOK WITH. USED IN *THAI* & **SOUTHEAST ASIAN** SOUPS, STIR FRYS, & CURRIES.

LEMON BASIL

A **WONDERFUL** LEMON SCENT. *Beautiful,* COMPACT PLANTS. USED iN INDONESIAN, *Thai* & NORTH AFRICAN SOUPS & SALADS. THE DRIED SEEDS ARE EVEN USED iN *sweet THAI* desserts.

FOODIE GARDEN

TRIPPLE XXX PEPPERS

3 HOT PEPPERS to enliven your GARDEN and heat up your COOKING.

HABANERO.
40 TIMES HOTTER than Jalepeno! small but MIGHTY Matures in 90→100 DAYS to a Lovely YELLOW-GOLD colour. For SUPER HOT SALSA, SPICY MOLES & HOMEMADE

PEPPERONCINI
A Tad Hot but with a FRESH FLAVOUR. Ripens to light green colour. 4" long WRINKLY fruit. A good pepper to PICKLE, TOSS in a SALAD or ON TOP of a PIZZA.

CAYENNE.
This SLIM Hottie is a CLASSIC HOT PEPPER. LONG, THIN 5" RED BEAUTIES, Curled & Twisted RED HOT TASTE Great for STIR FRIES or CURRIES Easy to dry & STORAGE for use Later.

FOODIE GARDEN

TOMATO TAKEOVER

NATIVE TO SOUTH AMERICA, THE TOMATO TOOK OVER THE WORLD AFTER THE CONQUISTADORES BROUGHT THEM BACK TO Spain. NOW THE ROCK OF Italian & Mediterranean COOKING, the KING of the SALAD & Topper of Burgers.

BLACK KRIM
BiG, BURLY 3 to 4 inch DARK MAHOGANY-COLOR BEEFSTEAK, TOMATOES. GETS DARKEST in HOT WEATHER! RICH FLAVOUR WITH SLIGHT SALTINESS.

Green Zebra
STRIPED SALAD TOMATO, WHICH RIPENS TO AN AMBER-GREEN WITH DARK STRIPES. THESE 3 ounce Bi-COLOUR BEAUTIES TASTE Sweet & ZINGY.

ROMA
THE Classic Italian RED SAUCE TOMATO. COMPACT VINES WITH 3 inch RED PEAR-SHAPED FRUITS. SOLID WITH FEW SEEDS. THAT'S AMORE!

TRUE

Tasty

Cinn-O-mite!

CiNNAMON BASiL

BLiSTERING

PEPS

BLAZING

FOODIE GARDEN

Dark -N- Lovely

BLACK CHERRY TOMATOES

BLACK CHERRY TOMATOES

The DARKEST in cherry tomatoes packing a RICH, AWESOME tomato flavour. Pop these little beauties on top your salad or directly into your mouth.

Native to SE Asia.

FRAGRANT, *SWEET*,

CiNNAMiC! NATUP

Naturally contains cinnamate the same thing thing that gives

Mainly founded & used in
not to get burned!
the Car ibean islands,

these blistering hot balls of

fire ripen to blazing
orange color. Be careful

Holland says, "The brief was pretty open. For the color sketches, I was running with silk screen inspiration in the vein of Globe Poster or Hatch Show Print—bright, simple color separations, bold type. In the end, I'm happy we went one-color, as the product was allowed to become its own thing, not a derivative."

LIMITED-EDITION FILM AND PACKAGING
Sagmeister & Walsh

Sagmeister & Walsh is a New York City–based design firm that creates identities, commercials, websites, apps, films, books, and objects for clients and audiences as well as for themselves.

PRODUCT: Limited-edition film and packaging

DESIGN FIRM: Sagmeister & Walsh

CREATIVE DIRECTOR: Stefan Sagmeister

ART DIRECTOR: Santiago Carrasquilla

DESIGNERS: Santiago Carrasquilla, Christian Widlic, Esther Li, and Thorbjørn Gudnason

HAND LETTERER: Stefan Sagmeister

CERAMIC PRODUCTION: Janine Sopp

BOX PRODUCTION: South Side Design and Building

MEDIUM: Paint

COUNTRY: United States

PROJECT DESCRIPTION: For three of their most recent typographic films (*If I Don't Ask, I Won't Get*; *Be More Flexible*; and *Now Is Better*), shown as part of the exhibition *The Happy Show*, Sagmeister & Walsh created limited-edition boxes (editions of ten), each containing an earthenware USB drive (specific to each film), a Blu-ray disc, and a certificate of authenticity. Handwriting played a central role within the exhibition. It made sense to extend it to these packaging elements. The type used in the project is part of Stefan Sagmeister's ongoing exploration of the motif, "Things I have learned in my life so far."

Each box is literally lettered by hand, making each a more collectable item. Is there any other reason you did that?

STEFAN SAGMEISTER: As the edition was only ten each, going through the trouble of having films made and silk-screening them would have been just as much work as sitting down and handwriting them all. And since I had handwritten all the text on the walls for the various stops of the exhibit already (Philadelphia, Toronto, Los Angeles, Chicago, Paris, Vancouver, and Vienna), it seemed in keeping with the overall spirit of the project.

Custom lettering and type play a role in your work in general. Talk about that (reasons, challenges, advantages, disadvantages). Why not Helvetica?

When I was fifteen, I was writing for a small local magazine and discovered that I liked creating the layout better than writing. We were setting headlines with Letraset sheets donated by friendly design studios, and as they invariably all had the letter *e* missing, it was easier to write a headline by hand than to reconstruct the missing *e*. That's where my love for handwritten typography stems from.

Much later we created the series "Things I have learned in my life so far." My grandfather was educated in sign painting and I grew up with many of his pieces of wisdom around the house: traditional calligraphy carefully applied in gold leaf on painstakingly carved wooden panels. I am following his tradition with these typographic works. They are all part of a list I found in my diary under the title "Things I have learned in my life so far."

The message is always very clear and straightforward, the typography much more ambiguous and open for interpretation. I found that by utilizing an open typographic approach combined with a clear message, many viewers have an easier time relating to the work. We do employ various typographic

Sagmeister's own handwriting was chosen to evoke the simple, clear messages of this project's theme: Things I have learned in my life so far. Because this series of products was limited to a total of thirty, it was deemed easier to letter each by hand.

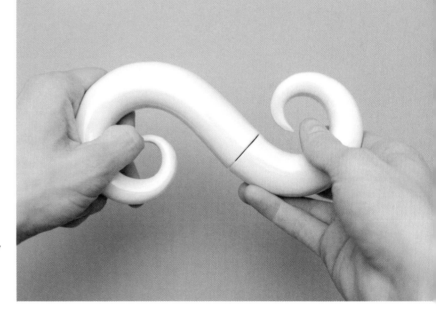

strategies from one project to another (within the series). Some are influenced by the environments in which they take place, some by outside people, some by personal experiences.

You've remarked on the use of handwritten lyrics on your Lou Reed poster, saying it's "very personal." Can your handwriting remain personal across a range of projects (like the movie boxes)? Does it become impersonal if repeated too often, or has it become your signature?

Everything will eventually lose its meaning through repetition. I remember promising myself ten years ago to not use my handwriting again, for fear that I had done it too often already. Since then, a project with a compelling reason for breaking that promise seems to come along every so often.

On future work, Sagmeister says, "Recently, I copied a page from an eighteenth-century Turkish Quran I had seen in the Islamic Arts Museum Malaysia in Kuala Lumpur. I hope this has an influence on our work, as Western typography has rarely reached the absolute pinnacle of total gorgeousness as it did in Islamic culture (as they were not allowed to show any imagery, all the creative juice went into type and ornament)."

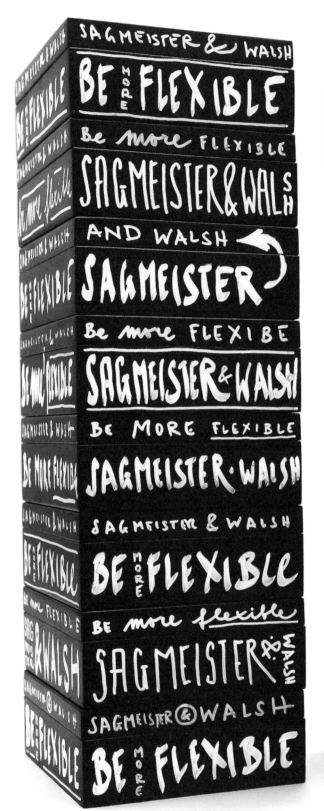

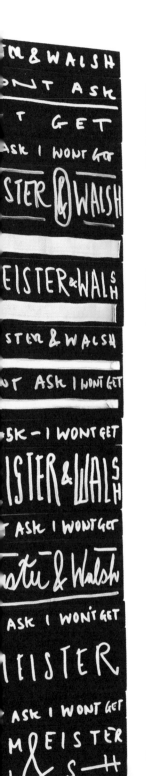

Sagmeister's films include the vernacular typography that has made him a much-admired figure in graphic design.

Studies for the boxes, lettering, and various ceramic objects used to house
USB drives containing typographic films

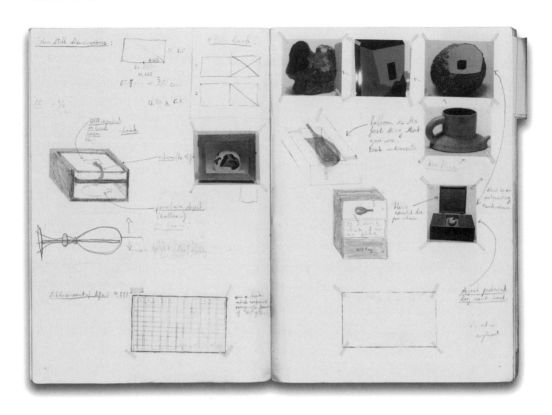

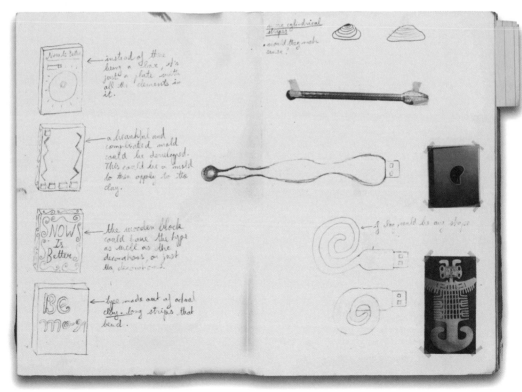

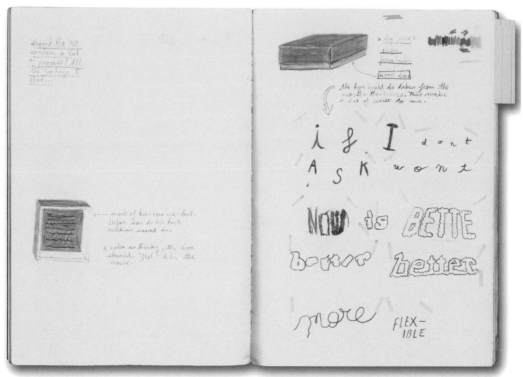

MANIKAY WINE
Cult Partners

Cult Partners is an award-winning San Francisco–based design firm that creates thoughtful brand experiences. With a specialization in packaging and identity programs, they offer solutions tailored to each of their clients' individual needs, rather than pushing their own aesthetic or agenda. Cult Partners's goal is to create lasting and memorable solutions for their customers that will help differentiate them within the marketplace.

Cult Partners designs packaging, identities, structures, marketing collateral, point-of-sale materials, and websites, and offers logo development, brand strategy, and naming services.

PRODUCT: Manikay Wine
CLIENT: Winery Exchange
DESIGN FIRM: Cult Partners
ART DIRECTORS/DESIGNERS/
ILLUSTRATORS: Jeff Hester and
Sara Golzari
HAND LETTERER: Jeff Hester
PHOTOGRAPHER: Ronnie Poon
MEDIUM: Sharpie, India ink,
and digital
COUNTRY: United States

PROJECT DESCRIPTION: The Aboriginal people of Australia have been singing manikay for centuries. Manikay are a series of songs, passed down through generations from ancestors, containing knowledge essential to their way of life. They often celebrate the natural rhythms of the land, and the plants, animals, and people that inhabit it. For Manikay Wine, Cult Partners hand-drew and silk-screened a dot pattern on the wine bottle to symbolize the musical rhythm of the manikay and the visual rhythm the hillside vineyard rows take on before harvest in the Barossa Valley of Australia.

Positioned to appeal to affluent millennials looking for a nontraditional wine offering, the resulting look of the Manikay label is distinctly Aborignial Australian, but with a slightly abstract and modern feel.

How were you able to get away with using such small type on a commercial product? Please share your secret!

JEFF HESTER: Wine is a unique product—it is CPG (consumer packaged goods), but people expect it to be a little more artistic and refined. Clients do not expect type to be an inch tall, bold, and have a drop shadow. An Opus One wine bottle would not feel the same if the branding was an inch tall and bold.

That said, the Manikay branding is a little on the small side. The client and I thought the overall package was bold and different enough to compel the consumer to pick up the package and read the brand name.

What makes the packaging distinctly Aboriginal Australian?

We researched aboriginal paintings, drawings, tattoos, and decorations. Dot patterns are a common theme. After looking at all the research, we tried to distill the images into something very modern and simple. We combined the dot pattern with the topo maps, photographs of vineyard rows, and images of soil/sedimentary strata. All have rhythmic lines.

How are the dots representative of rhythm?

There are several layers of rhythm in the dots. There is a rhythm in the various sizes of dots, the dots are uniformly spaced, and the lines of dots are uniformly spaced.

Initially, Cult Partners wanted the dot pattern to go all the way up to the neck, but that was not possible on ACL (applied color label) printed bottles. The next best solution was to crop the dots on an angle to make it more organic. They ended up preferring that solution.

MANIKAY

20/12

MANIKAY

MANIKAY

2012

BAROSSA

SHIRAZ

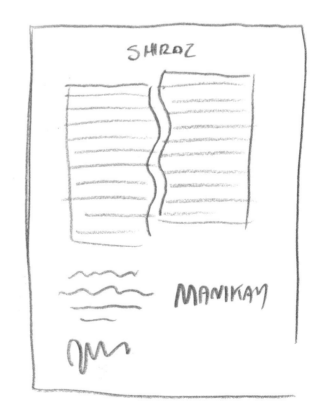

SHIRAZ

MANIKAY

MANIKAY

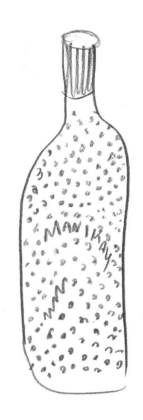

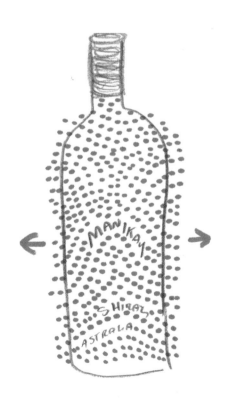

Hester's sketches include abstract rivers, layers of soil strata, and different Aboriginal Australian patterns. One even incorporated die cuts. "The die cut was elminated," Hester says, "because it probably would not hold up on a bottling line."

"We experimented with a bunch of different ways to draw the artwork," says Hester. "We tried Sharpies, ballpoint pens, and hand-carved bamboo pens, but we ultimately used matchsticks dipped in India ink."

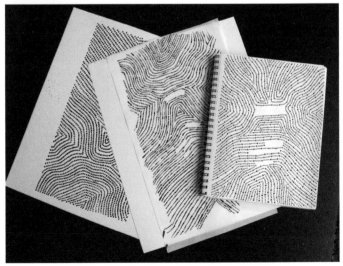

CHIPOTLE MEXICAN GRILL
Sequence

  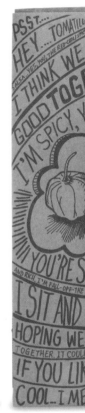

PRODUCT: Chipotle Mexican Grill
CLIENT: Chipotle Mexican Grill
DESIGN FIRM: Sequence
CREATIVE DIRECTORS: Anna Tou,
AJ Kiefer
ILLUSTRATOR/HAND LETTERER:
Anna Tou
COPYWRITER: AJ Kiefer
MEDIUM: Sharpie
COUNTRY: United States

Sequence is a creative development agency with offices in San Francisco and New York. A team of strategists, designers, and technologists, they create brands, digital products, and connected experiences that resonate with people and grow businesses. Their clients include iconic brands such as Chipotle, Chevron, Apple, Disney, Peet's Coffee & Tea, Food Network, Best Buy, and Lytro.

PROJECT DESCRIPTION: When Chipotle asked Sequence to revamp its packaging to present ideas around their slogan, "Food with Integrity," their answer was "passionate ramblings," a voice and illustration style that has become an iconic hallmark of the brand since its debut in 2010. The playful packaging got people reading while eating, and laughing while they were at it. Sequence painstakingly hand lettered every single design, populating bags, cups, napkins, walls, and even a few commuter trains with a mix of over forty unique stories that are effusive odes to the freshness and deliciousness of Chipotle's food.

How does the question of legibility come into play when working with hand-drawn lettering?
SEQUENCE: It's a really big deal! In this

packaging system, it was important that the copy come across clearly and communicate the core Chipotle Mexican Grill message—

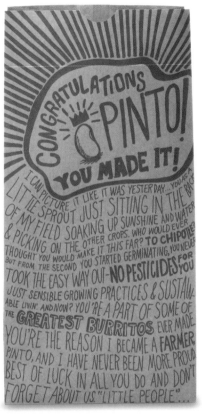

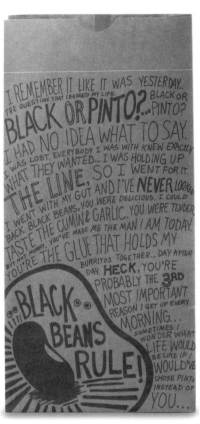

however, it also still needed to work as an aesthetic system and texture, even if a customer doesn't take the time to read the full story.

How much of this copy was provided vs. self-generated?

All of the copy for this packaging system was generated by Sequence, in close collaboration with Chipotle marketing.

Is it more "love note" or more "manifesto"?

This is a great question—the idea was to bring passion and humor to what could be construed as dry business statements about quality, sourcing, and sustainability. Customers are hungry to learn about why

Chipotle is different, but the brand tries to avoid being preachy or "downers" when talking about these important issues. The idea was to find that balance of information and entertainment.

Is this totally freehand, or are there pencils first that we can't see?

Most of the lettering was completed freehand with Sharpies, and very rarely a pencil was used for a preliminary outline—and even then, it was almost only ever for an illustration. The pieces were then scanned in and small "microedits" were completed in Photoshop. However, every single piece was hand sketched with no computer illustration.

Marker drawings were the core designs for the original six packaging pieces. These first six bags were the beginning of the entire system—growing to over forty different bags, cups, and boxes of varying sizes.

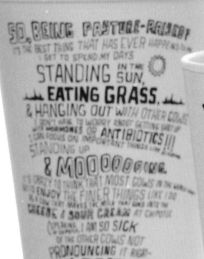

Cup 1 (left):

...A LITTLE PEPPER GUY— IF THAT'S THE GOOD FOR YOU MAN!
...MOR YOU ENDED UP AT CHIPOTLE! ...D A FUTURE WHEN I PICKED YOU OFF THE VINE. ...U WERE **SO RED & RIPE** ...ADY FOR THE WORK YOU HAD TO DO. ...ORGET THE LOOK ON YOUR FACE WHEN WE LAID YOU ...E SUN WITH ALL THE PECAN-WOOD SMOKE ...NG YOU ALL **HOT.** SOME OF THE OTHER PEPPERS WERE COMPLAINING ...FLAVOR YOU NEEDED TO MAKE IT ALL THE WAY **YOU NEVER** ...GAVE UP, YOU DID WHAT YOU HAD TO DO. NOW YOU'RE ...RT OF A PRETTY OLD-SCHOOL TRADITION, BUDDY ...EEN GROWING & SMOKING ...'S LIKE YOU FOR OVER **200** YEARS DOWN HERE ...UAHUA & IT'S ALWAYS NICE WHEN A PEPPER LIVES ...TO OUR STANDARDS. AND YOU'RE SURROUNDED ...L THOSE OTHER INGREDIENTS THAT ARE ...KY ENOUGH TO BE A PART OF CHIPOTLE'S ...OD— IT'S **TOUGH** COMPETITION ...EASON FOOD LIKE THAT. ...YOU, MY LITTLE PEPPER BUDDY, HAVE ARRIVED. SO JUST TAKE A MINUTE TO RELAX, BUD. ...NJOY THE **BURRITO** DREAM ...VE GOTTEN YOURSELF INTO... I CAN'T WAIT TO SEE WHAT YOU'RE UP TO NEXT, **JUAN**

Cup 2 (center):

THANK YOU FOR YOUR INTEREST IN BEING A PART OF CHIPOTLE **GUACAMOLE** YOU COULD GO AHEAD AND ANSWER THE FOLLOWING QUESTIONS WE CAN START OUR **APPLICATION PROCESS:** Q: **WHY DO YOU WANT TO BE A PART OF CHIPOTLE** A: **WELL FIRST OFF,** I HAVE ALWAYS BEEN A FAN OF YOUR GUACAMOLE, I LOVE THE IDEA THAT NOT ONLY **TASTES GOOD,** BUT IS ALSO MADE THE **RIGHT WAY** SEVERAL TIMES A DAY. I AM NO FAN OF SHORT CUTS. Q: **DO YOU WORK WELL** WITH OTHER INGREDIENTS? **ONIONS? JALAPEÑOS?** HOW ABOUT **LIME**?!!! A: I WORK REALLY WELL WITH ALL OF THOSE GUYS EVEN THOUGH LIME TENDS TO BE A LITTLE OVERBEARING IF YOU DON'T KEEP HIM IN CHECK. WE'VE WORKED TOGETHER BEFORE, THE IDEA OF GETTING BACK TO BASICS JUST US INGREDIENTS & SOMEONE WHO KNOWS HOW TO **SMASH** US UP JUST Q: **SHARE ONE STRENGTH** & ONE WEAKNESS A: STRENGTH... WELL, I'M **GREEN.** THAT'S TRENDY RIGHT? AS FOR A WEAKNESS, I TEND TO BE SLOW TO GET ACCLIMATED TO A NEW ENVIRONMENT. YOU COULD SAY I NEED SOME TIME TO **"RIPEN"** BEFORE I'M Q: **WHERE DO YOU SEE YOURSELF IN 5 YEARS** A: I'M AN **AVOCADO.** I DON'T THINK ABOUT ANYTHING THAT FAR OUT.

Cup 3 (right):

YOU CAN JUDGE ME IF YOU WANT TO, ...C BUT **I DON'T CARE** THAT LITTLE BIT OF SALSA & CHEESE **HANGING OUT** ...IT'S ALL **MINE.** ...MEET AWESOME PORK EVEN AND SOMETIMES IT'S CHICKEN OR STEAK... **SO FINE.** I GUESS YOU CAN GO AHEAD & CALL ME A **"FOIL LICKER"** GET THAT LAST HOLD OF THAT ONE LAST DELICIOUS **BITE** THAT'S EVER THINK OF THAT'S LIKE MY BURRITO IN A BANK THAT NEEDED A **BAILOUT**

Foodie manifestos sing the praises of Chipotle, and seek to provide customers with a balance of information and entertainment. Hand letterer Anna Tou's work can be seen across the entire Chipotle branding system, on everything from cups to walls.

MARIA'S PACKAGED GOODS
& COMMUNITY BAR
Franklyn

Franklyn likes to think of itself as the creative studio you've been searching for all your life, whether you realized it or not. Founded in 2012 by Michael Freimuth and Patrick Richardson, they're a Brooklyn-based studio of six, trained as graphic designers, illustrators, motionographers, strategists, and creative technologists. For those who may not know, their favorite color is green.

PRODUCT: Maria's Packaged Goods & Community Bar
CLIENT: Ed Marszewski, Maria's Founder
DESIGN FIRM: Franklyn
ART DIRECTOR/HAND LETTERER: Michael Freimuth
PHOTOGRAPHER: Marlene Rounds
MEDIUM: Ink and paint
COUNTRY: United States

PROJECT DESCRIPTION: Franklyn was asked to create an identity and packaging system for Maria's, a classic "slashy" (half liquor store/half bar) in Chicago's emergent creative neighborhood of Bridgeport. "Maria" is the name of the owner's mother, who had previously run the space under another name prior to handing it over to her son. Franklyn was asked to retain the warmth the owner felt for this family-run and neighborhood-friendly establishment, prompting an exploration of hand-drawn letterforms with an emphasis on the casual and familiar. The result is a mash-up of handmade and rubber-stamped goodness.

You do a lot of custom typefaces, which are often quite elegant. This project seems like the antithesis of those faces. Do the two styles/techniques/approaches overlap or inform each other at all? Does the OCD part of your brain (good for type design) fight against being fast and loose?
MICHAEL FREIMUTH: Actually I think it's the opposite. The OCD part feels liberated after being ignored for a while. Perhaps our current body of work at Franklyn leans toward the (in your words) elegant side of typography, but historically, I'm more comfortable in the fast and loose realm. In a related manner, I think of the two "styles" (elegant versus quick and dirty) as being just two expressions on a much larger spectrum—not necessarily binary opposites, simply different tools in an enormous repertoire.

Explain the paper wrappers. How will they be used? Is there a label underneath?
The beer and liquor from Maria's used to be sold and then wrapped in old newspaper or paper—whatever happened to be around in the store. We felt retaining that material via the use of blank newsprint was a simple, cost-effective way to stay true to Maria, while allowing for personalization, quirkiness, and the occasional touch of wit. The wrapping is for both the existing beer and liquor brands sold by the store and for a healthy amount of craft-brewed beer that Maria's has begun to produce themselves. These craft beer bottles don't have conventional labels, per se. Instead, the bottles are first painted black or white and are then either left enigmatically blank or hand-lettered by Maria's staff.

OPPOSITE: *Maria's is a hybrid bar/liquor store that developed its own line of beverages. The finished package is ready for consumption by thirsty patrons.*

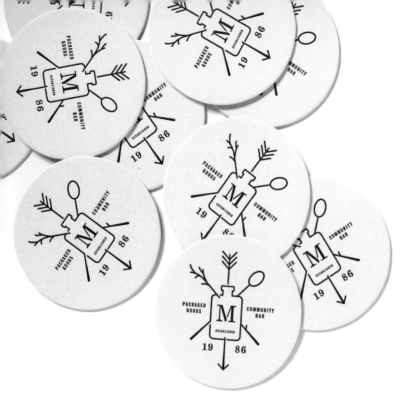

The DIY-style packaging draws its inspiration from the bar's history of wrapping liquor to go in miscellaneous bits of found paper.

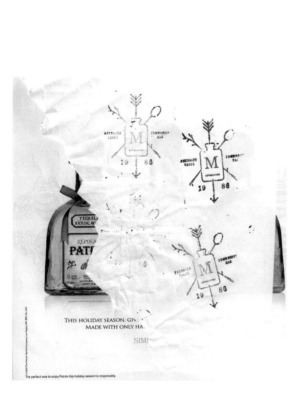

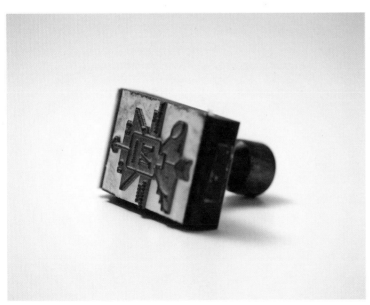

Maria's LIQUOR BAR

Maria's

Maria's

Maria's

Maria's

Maria's

Maria's

MARIA'S

MARIAS

MARIA'S

MARIA'S

MARIA'S

Maria's

Franklyn's variety of sketches riff on casual napkin scrawls and unselfconscious writing styles to achieve an overall down-and-dirty bar vibe.

PACKAGED GOODS
& COCKTAIL LOUNGE

Packaged Goods
& Cocktail Lounge

MARIA'S

MARIA'S
PACKAGED GOODS
& COCKTAIL
LOUNGE

MARIA'S

Maria's

packaged
goods & cocktail
lounge

PACKAGED
GOODS &
COCKTAIL
LOUNGE

PACKAGED
GOODS &
COCKTAIL
LOUNGE

Marias

Est. & & &
& & &

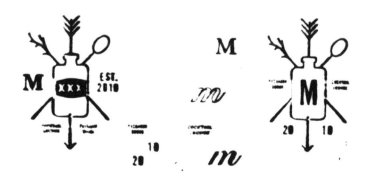

LOVE IN A CUP
Elmwood

Elmwood began as a small, local agency in 1977, and while they now have studios in Leeds, London, New York, Singapore, Hong Kong, and Melbourne, they're still very much independent.

Elmwood has been accredited a silver standard by *Investors in People* and named the *Sunday Times*' "Best 100 Companies to Work For" in 2008 and 2011. They provide everything from insight and strategy to creative and design, and work with a long list of sectors including retail, government, sports, media, FMCG (fast moving consumer goods), B2B (business to business), and more. Their joint ventures also mean they can offer services in sustainability consulting, industrial design, and environmental design.

PRODUCT: Love in a Cup
CLIENT: Wm Morrison Supermarket PLC
DESIGN FIRM: Elmwood
CREATIVE DIRECTOR: Martyn Hayes
DESIGNER: Simon Thorpe
HAND LETTERER/ILLUSTRATOR: Kate Pankhurst
COPYWRITER: Liz O'Connor
MEDIUM: Pen and ink
COUNTRY: United Kingdom

PROJECT DESCRIPTION: Love in a Cup was a fun and unique Valentine's Day gift intended to capture a moment between loved ones, shared over a cup of tea, much like a handwritten card. The hand-drawn design of the packaging evoked a personal love note.

What is the connection between romance, tea, and hand-drawn typography?
MARTYN HAYES: Sometimes it's the simplest things—writing a little note of affection or making a drink for someone creates a tender moment between loved ones, from new couples to older married folks. The connection between the two created the design solution.

Were the different tag messages the client's idea or Elmwood's?
The client thought there was an opportunity to use the tags. Elmwood worked with the copywriter to develop five different romantic messages on the tags, showing your loved one you care every time you put the kettle on.

Was the typography created in-house or did you work with an outside lettering artist?
The typographic style and direction were developed by the designer, but the final illustration was created by the illustrator Kate Pankhurst.

What other options were explored in the process of arriving at this solution?
We presented a number of concepts to the client, ranging from very pictorial illustration ideas to much more typographically led routes. The type solutions were felt to have more warmth and the same sentiment as a Valentine's message in a card.

M

'LOVE' in a cup

FORGET FLOWERS AND CHOCOLATES~SOMETIMES IT'S THE *little* THINGS THAT SAY 'I love You'. OUR **LOVE in a CUP** TEA BAGS COME WITH FIVE DIFFERENT ROMANTIC MESSAGES ON THE TAGS, SO YOU CAN SHOW YOUR LOVED ONE YOU CARE EVERY TIME YOU PUT THE KETTLE ON...

20 RED LABEL TEA BAGS WITH LOVEHEART TAGS

YOU STIR ME UP

Because nothing says 'I LOVE YOU' like a cup of tea in bed

M

'LOVE' in a cup

20 RED LABEL TEA BAGS WITH LOVEHEART TAGS

YOU STIR ME UP

Because nothing says 'I LOVE YOU' like a cup of tea in bed

hey sugar!

You're Sweet

HOT STUFF

SQUEEZE me.

YOU STIR ME UP

Romantic messages are the secret ingredient in Love in a Cup's hand-drawn tea bag Valentine's gift.

BAGS OF LOVE

THE CUPPA WITH a heart

BAGS OF LOVE

BECAUSE NOTHING SAYS I LOVE YOU LIKE A CUP OF TEA IN BED

FORGET FLOWERS AND CHOCOLATES ~ SOMETIMES IT'S THE LITTLE THINGS THAT SAY 'I LOVE YOU'. OUR **BAGS OF LOVE** TEA BAGS COME WITH FIVE DIFFERENT ROMANTIC MESSAGES ON THE TAGS SO YOU CAN SHOW YOUR LOVED ONE YOU CARE EVERY TIME YOU PUT THE KETTLE ON

say you care with...

BAGS OF LOVE

THE CUPPA WITH a heart

LOVE
in a cup

20 RED LABEL TEA BAGS WITH LOVEHEART TAGS

Because nothing says 'I LOVE YOU' like a cup of tea in bed.

FORGET FLOWERS AND CHOCOLATES ~ SOMETIMES IT'S THE *little* THINGS THAT SAY 'I love you'. OUR **LOVE in a cup** TEA BAGS COME WITH FIVE DIFFERENT ROMANTIC MESSAGES ON THE TAGS SO YOU CAN SHOW YOUR LOVED ONE YOU CARE EVERY TIME YOU PUT **THE KETTLE ON...**

BAGS of LOVE

BECAUSE NOTHING SAYS I LOVE YOU LIKE A CUP OF TEA IN BED

20 TEA BAGS WITH 5 ROMANTIC MESSAGES

LOVE in a cup

50 BLACK TEA BAGS WITH 5 ROMANTIC MESSAGES

BECAUSE NOTHING SAYS I LOVE YOU LIKE A CUP OF TEA IN BED

LOVE in a cup

BECAUSE NOTHING SAYS I LOVE YOU LIKE A CUP OF TEA IN BED

20 TEA BAGS WITH 5 ROMANTIC MESSAGES

LOVE in a cup

Show you care with THE CUPPA WITH A HEART

20 TEA BAGS WITH 5 ROMANTIC MESSAGES

LOVE in a cup

Show you care with THE CUPPA WITH A HEART

20 TEA BAGS WITH 5 ROMANTIC MESSAGES

JAMESON WHISKEY ST. PATRICK'S DAY 2014
Dermot Flynn

O riginally from Dublin, Dermot Flynn is a London-based designer and illustrator. He has worked for various clients, such as Pentagram, Adidas, Vodafone, *New York* magazine, Condé Nast, Dentsu Japan, Winkreative, *GQ*, the Guardian, BMW, Francfranc, SWISS, Nickelodeon, and the *Irish Times*.

PRODUCT: Jameson Whiskey
St. Patrick's Day 2014
"The Spirit of Dublin"
CLIENT: Jameson
AGENCY: Design Bridge
ART DIRECTOR: Asa Cook
DESIGNER/HAND LETTERER/
ILLUSTRATOR: Dermot Flynn
MEDIUM: Pencil, graphite, pen, ink, Sharpie, acrylic, MacBook, and scanner
COUNTRY: Republic of Ireland

PROJECT DESCRIPTION: In April 2012, Jameson Irish Whiskey contacted Flynn about creating a pitch for the new label design for their 2014 limited-edition St. Patrick's Day bottle. Jameson works with a different Irish artist each year to create the bottle's label. The brief's title was "The Spirit of Dublin." The aim of the bottle design was to evoke the sights, sounds, and atmosphere of a St. Patrick's Day celebration in the Irish capital. Flynn created a variety of designs and concepts, one of which led to the final design. He then worked with the London consultancy Design Bridge, which helped with the label's final execution and production.

Your work seems primarily illustration based. How did you first become interested in lettering?
DERMOT FLYNN: When I studied visual communication at the National College of Art and Design in the mid 1990s, there was a great emphasis on typography and familiarizing oneself with letterforms and type etiquette. We also had an array of visiting lecturers who would run short, one- or two-week projects, one of which was a calligraphy workshop led by renowned Irish calligrapher Frances Breen. An initial exercise was to make a calligraphic mark, with the aim being to relax and not think too much when drawing shapes and letterforms—a welcome change from the tight, precise work we had been doing with typefaces, such as Helvetica, Baskerville, and Futura. After this, I started looking more and more at hand-drawn type when designing—in particular the work of designers like Paul Rand, Saul Bass, and Alan Fletcher—and it just developed from there. As I became more and more comfortable with computers and scanning and manipulating images, hand-drawn lettering began to play a greater part in my work.

How do you think about the interaction between letter and image? Or as an illustrator, how do you think about type?
I guess I have a pretty intuitive approach.

I try not to labor a point too much when creating accompanying type with an image. I worked for Carton LeVert for a period after finishing college, where I really learned the ins and outs of typography, and I remember Libby Carton saying that there was a time and a place for everything. It varies from job to job. Ideally one won't overpower the other.

Where did all the humorous sayings come from? Is it supposed to be (bathroom) graffiti?
Ha! Well—various people and places! My mother is from the south and my father is from the northwest of Ireland—and they both had a whole series of sayings that were trotted out on various occasions. I also grew up in the north side of Dublin, which has a very funny, irreverent, and unique sense of humor. The novels of Roddy Doyle—*The Commitments* and *The Snapper*—were also an influence. Also, overheardindublin.com— it's a website where people log funny things they hear Dubliners say in the city's streets, pubs, and shops. Some of them, as you may imagine, are not exactly PG-rated! Another factor I guess was the graffiti in the National College of Art and Design toilets while I was studying there. Visual Communication students would put as much effort and imagination into their cubicle wall postings as they would with their studio projects and assignments!

The packaging for Jameson's 2014 limited-edition St. Patrick's Day bottle uses loose, energetic lettering to capture the "Spirit of Dublin" with the help of familiar phrases overheard on the streets of the city.

*The MO for these initial sketches was
to shoot first and ask questions later.
Then edit down for the final project.*

GOWONOWADAT!
STORY, HORSE? I'M HERE! CAN YOU
NOT SEE ME? I'M
?
SCAOIL AMACHÉ!
And I lost my ♥ to a GALWAY GIRL
YOUNG FELLA!
FINE bit of Stuff
WEARING GREEN....
RAPID! ME ♥ UL' SEG ♥ TIA
GERRUP outta dat Serious?
15 Euro for Razors? Grand
WERE they FORGED in the FIRES of HADES?
Stop the Lights
Are ya serious?
DEADLY!
the Head on that and the PRICE of CABBAGE....
I will in me Granny
Ah Lord...
A NECK like a JOCKEY's..
GEHOUT Ò DAT GAR DEN

G'NOSH
Mystery Ltd.

PRODUCT: G'NOSH
CLIENT: G'NOSH Ltd.
DESIGN FIRM: Mystery Ltd.
CREATIVE DIRECTOR/DESIGNER/
HAND LETTERER: Phil Seddon
PHOTOGRAPHER: Dan Einzig
MEDIUM: Acrylic white paint on
uncoated board
COUNTRY: United Kingdom

Mystery Ltd. is a London-based specialty brand design studio that creates integrated work for the hospitality and food and beverage sectors. This ranges from the creation of restaurant and bar interiors and brand environments to food and drink branding and packaging. The agency makes full use of traditional and modern production processes, such as creating graphic installations designed to build intrigue and give brands an authentic, rich, human touch through sign writing and prop making.

PROJECT DESCRIPTION: Mystery was approached by passionate "foodie" entrepreneur Charlotte Knight to create G'NOSH, a brand for her new line of gourmet dips that would shake up the UK's dull supermarket offerings of hummus and mayo-heavy party packs. The brief took the project from strategy and naming through to final packaging and digital design. Mystery's task was to create a memorable brand with real personality in order to differentiate the products on the shelf and online.

You've had tremendous success with the brand. Why do you think consumers have connected so readily with it?

PHIL SEDDON: With the same team working on all stages of the project, everybody involved had a really clear idea of what the brand represented. The brand's mantra, "We believe in the power of sharing," was a message that the consumers took to heart, and with founder Charlotte active in mentoring and partnering with other food brands, that ethos permeated the whole business. I think that means that consumers believe in the brand, and they have responded well to the hand-painted lettering that reinforces this human, noncorporate approach. We were inspired by hand-written campaign placards that matched the social and proactive ambition of the business.

Was the type painted freestyle or did you first design it on the computer?

The idea for the typography was always to keep it completely freestyle. We wanted it to feel as natural as possible. I try not to labor over the execution of the type so that it doesn't look too contrived. I often have a rough idea of the composition of the pack in mind so I know when I should curve the type around a certain element, but generally the final composition happens at the final stage so I just crack open the paints and have fun. Getting away from the all-pervading digital screen is extremely therapeutic!

Why was hand-painted type appropriate for this project?

The brand called for a human, authentic, campaigning tone of voice—hand painting was the ideal way of representing this visually. Really, it's the interaction between

The dips that established the growing G'NOSH gourmet food brand

the type and the ingredients that gives the brand its individuality—so the shadows that are cast onto the type are what bring the voice of the brand and the quality of its gourmet ingredients together. This also helps communicate that the recipes are refined by hand rather than in the factory. Hand painting gives the brand warmth and a sense of attention to detail that a digitized typeface just can't emulate.

When working with the client on a project that is 100 percent handcrafted, how do you manage edits and changes, especially to the type?
It's important to us that the brands we design are practical as well as visually appealing, so it was key that we could demonstrate a way of rolling this concept out without losing the authenticity. Happily, utilizing simple technology in the form of a scanner and Adobe Photoshop enabled us to compose the final image and edit type and layout easily, while still using hand lettering and fantastic food photography. Thankfully, as the styling of the hand-rendered type is quite spontaneous, we aren't in a position where we have to repaint it over and over again, making edits fairly painless. As the brand grows, we are constantly looking for ways of refining the process to make it even more flexible without losing our handcrafted feel.

G'NOSH typography is hand painted onto black cards before being scanned and isolated in Adobe Photoshop, where the final brand environment is composed.

"A well-defined brand platform," says Seddon, Mystery's senior brand designer, *"based around 'sharing' underpinned a strong hand-painted look which, when combined with strong food photography in installations, allowed us to create a tactile brand world that works as well on-screen as it does on-pack."*

HEMA

SOGOOD

Having earned more than twenty years of international design experience at various agencies, Ron van der Vlugt, Rob Verhaart, and Edwin Visser founded SOGOOD in the Netherlands in 2008. SOGOOD is a small, dynamic design agency dedicated to increasing the commercial clout of their clients.

PRODUCT: HEMA
DESIGN FIRM: SOGOOD
ART DIRECTORS/ILLUSTRATORS/
PHOTOGRAPHERS: Ron van der
Vlugt and Rob Verhaart
HAND LETTERER: Rob Verhaart
CLIENT: HEMA
MEDIUM: Watercolor
COUNTRY: The Netherlands

PROJECT DESCRIPTION: HEMA is a Dutch private-brand retail chain store featuring both food and non-food items. It has more than 650 branches in Europe, in countries including the Netherlands, Belgium, Germany, France, and Luxemburg.

Established in 1926, HEMA has been the sole carrier of its own uniquely branded products. In the past, every product category had a different design style. Now, there is one look and feel, based on HEMA's principle of "exceptional simplicity." This design was developed for the more than four hundred SKUs of HEMA's food department.

The hand-drawn watercolors on a white background (sometimes in combination with photography) reflect HEMA's key values: natural ingredients, honesty, purity, freshness, quality, and deliciousness. In addition, the humor and whimsical aspects incorporated in the designs are supposed to elicit an optimistic, happy mood, and a heightened sense of enjoyment.

Were there many rounds of sketches before you arrived at the idea of hand-drawn type? How did you convince the client to buy in to your hand-painted type concept?
SOGOOD: We made five different concepts in close cooperation with HEMA for the new packaging assignment. From the beginning, both HEMA and SOGOOD were in favor of these hand-painted concepts, because they best reflect HEMA's key values.

How difficult is it to maintain brand consistency across a four-hundred-product range?
There are some differences between the different product groups. For example, the Take Away products received large hand-drawn lettering for their product descriptions. But overall there is consistency in that there is a white background and a simple watercolor, sometimes combined with a simple product shot. For each new product category, we develop a new concept. This concept will be translated to all SKUs within the category. If possible, each product gets its own individual character.

How is such a massive rebranding distributed among the designers in your studio?
Not everything was done in one week! We have been working for HEMA for almost two years now, going from one project to the other.

We have learned that not every designer is able to make HEMA-style watercolors, so this must be planned carefully.

How has the hand-drawn type trend played itself out in the Netherlands?
When we started there was no trend of using watercolors. Of course there were other types of hand-drawn illustrations in the market, but these were not that brand specific.

Nowadays, if anyone should use watercolor drawings (especially for food packaging) in the Netherlands, they would be considered copycats of the HEMA style.

Loose brush lettering, done in a bright watercolor palette,
helps keep this extensive product line fresh and simple.

AMERICAN COOKIE

American style cookie
Triple chocolate
Met drie soorten chocolade
Aux trois chocolats
Mit drei Sorten Schokolade

UTZ Certified cocoa

CAKE CHOCO SWIRL

Met de rijke
smaak van chocolade

AMERICAN COOKIE

American style cookie
White chocolate and raspberry
Witte chocolade en frambozen
Chocolat blanc et framboises
Weiße Schokolade und Himbeeren

UTZ Certified cocoa

CAKE LEMON FRESH

Met frisse citroensmaak

MUSTARD

Mosterd
Moutarde
Senf
Mosterd
Moutarde

Peper
Pepper
Poivre
Pfeffer
Poivre

MAYO

Fritessaus
Sauce a frites
Fritessauce
Fritessaus
Sauce a frites

KETCHUP

Tomatenketchup
Ketchup aux tomates
Tomatenketchup
Tomatenketchup
Ketchup aux tomates
Tomatenketchup

Zout
Salt
Sel
Salz
Sel

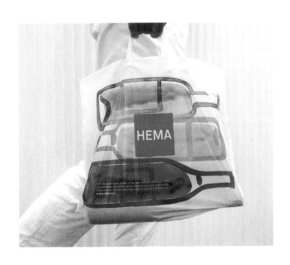

Each letter of the alphabet was repeated and then edited down to a selection used for final compositing. The briskly executed letterforms were then combined with lively, free-flowing illustrations.

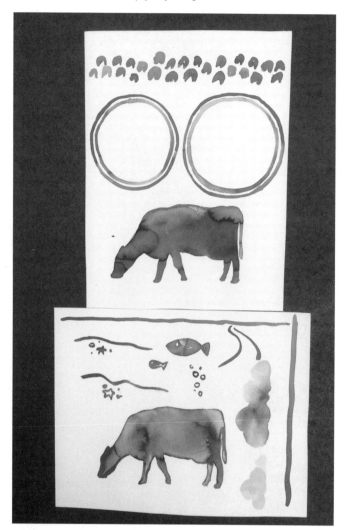

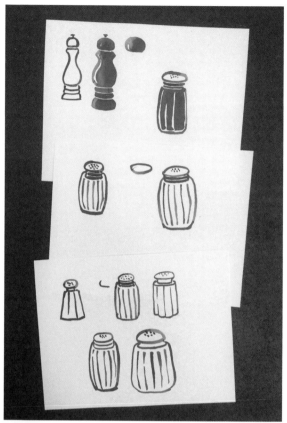

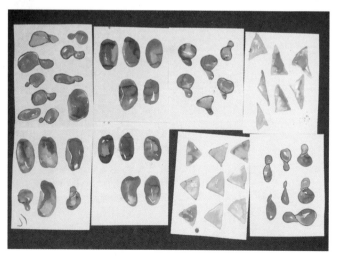

G G G G G G

C C C C C C C

B B B B B B

B H H H H H

J J J J J J J

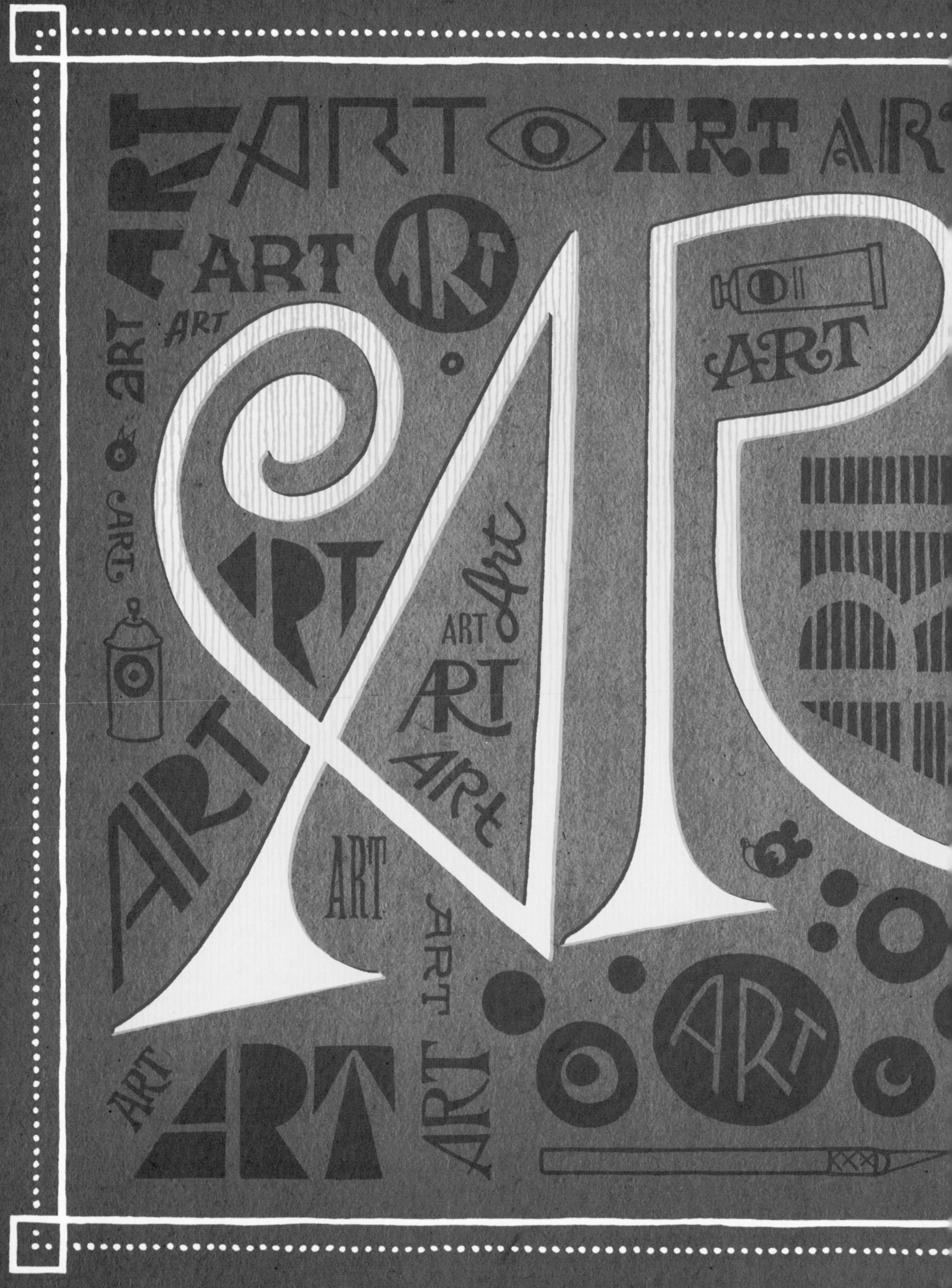
ART
ART
ART
ART
ART
ART
ART
ART
ART
ART
ART
Art
ART
ART
Art
ART
ART
ART
ART
ART
ART

It's been said that illustrators make great designers because they treat type as an image; with type and image rendered by the same hand, the packaging takes on a unique sense of cohesion and completeness. Here's packaging that's got character and whimsy, and, in some cases, a nostalgic flavor. Perhaps some of the magic is due to the almost anthropomorphic liveliness of the line work. Art and letterform mingle and dance, at times blurring into one.

BURNT SUGAR
d.studio

TITLE: Burnt Sugar

CLIENT: Burnt Sugar

DESIGN FIRM: d.studio

ART DIRECTORS/DESIGNERS:
Phil Curl and Wes Anson

ILLUSTRATORS: Gemma Correll,
Jess Wilson, Nick Deakin,
Kate Sutton, and Rudi de Wet

MEDIUM: Pen and ink

COUNTRY: United Kingdom

Wes Anson and Phil Curl of d.studio are brand designers based in the beautiful village of London, England. They work on large- and small-scale projects from clients across the globe, as well as closer to home. They love working with energetic emerging companies, matching their ambitions and sharing their enthusiasm to help brands achieve what they set out to do, and more. Their approach is to look at the bigger picture, creating narratives that run through brands from top to bottom.

PROJECT DESCRIPTION: Burnt Sugar is a boutique brand of fudge with humble beginnings. Started by Justine Cather, who nabbed her mother's recipe and took it to market (London's famous Borough Market to be more precise), the brand was small, but word soon got out and Burnt Sugar's popularity soared. Working closely with Cather, d.studio developed the "every one is different" concept, which celebrates not just the perfectly irregular shapes of taste bud–tingling bits of fudge, but also all the wonderful varieties of foodie folk who like to indulge in her confections.

The first thing they did was turn the pack around and tweak the flap to give the impression it was sealed with a sticker. This made it look like the traditional bags in which loose fudge is sold at Borough Market and instantly communicated the roots of the brand. They then commissioned five illustrators to doodle all over the packs so each product had its own individual personality. Not a single font was used on any of the packs—every element was meticulously hand rendered. D.studio also added a little surprise under the flap of each pack for the customers—or "fun-loving foodies," as Cather prefers to call them—to discover.

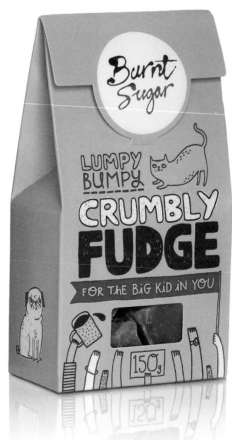

How did the client's brief evolve into what you delivered?
D.STUDIO: The initial brief was extremely open, so we were largely responsible for defining the brief from the outset. The biggest evolution (and leap of faith by the client) was going from the initial concept to the final packs, as a large amount of imagination was required to envision what the five packs would look like with a different illustrator's style on each one, which the client didn't see at first.

What was the process of rendering this type like?
It varied from pack to pack. Working with five different illustrators brought its own set of challenges. Some nailed it pretty much the first time, others required a few rounds of exploration. Our general approach was to set certain parameters that ensured the packs worked as a range but without limiting the creative expression of each illustrator—

we were keen for them to bring their own unique style and personality to the project.

Did the client request hand-drawn typography?
Not specifically, but Justine is a huge fan of illustration, so it didn't take much to twist her arm!

Do you see a trend of hand lettering taking place in the UK?
It's everywhere—from boutique brands to High Street private labels! But we like to think we got in before it was a "thing"— bandwagon jumping is banned here at d.studio!

A quirky variety of styles is used to reflect the "every one is different" concept, representing both the perfectly irregular pieces of fudge, and the lively mix of foodie fans who consume it.

Sketches for the sea salt caramel fudge package reflect a variety of fun concepts. The lettering styles complement the intricate handwork on the back side of the package, where even the nutritional informtation is hand lettered.

FROM THE START, WE'VE TAKEN OUR SWEET MAKING SERIOUSLY (ALTHOUGH NOT IN A FROWNY WAY - WE'RE STILL SMILEY FOODIES). WITH A LOT OF TRIAL & ERROR WE LEARNED EXACTLY THE RIGHT TEMPERATURE FOR MELTY CARAMEL CHEWINESS, AND WHY COMBINING THREE DIFFERENT TYPES OF SUGAR MAKES OUR CARAMELS TASTE EVEN BETTER. HUGGING THEM WITH SILKY, DARK CHOCOLATE GIVES THE PERFECT FINISH TO THESE SQUIDGY DELIGHTS

INGREDIENTS

CARAMEL (67%) [GLUCOSE SYRUP, CREAM, GOLDEN UNREFINED SUGAR, DEMERERA SUGAR, BUTTER, MUSCOVADO SUGAR, WATER, GOLDEN SYRUP, TREACLE, BALSAMIC VINEGAR, SEA SALT, EMULSIFIER: SOYA LECITHIN] DARK CHOCOLATE (33%) [COCOA MASS, SUGAR, COCOA BUTTER, EMULSIFIER: SOYA LECITHIN FLAVOURING] CONTAINS COCOA SOLIDS 55% MINIMIMUM

NUTRITIONAL INFO

TYPICAL VALUES PER 100G

ENERGY	1826 KJ
	433 kcal
PROTEIN	2.3G
CARBOHYDRATE	53.4G
OF WHICH SUGARS	23.7G
FAT	23.6G
OF WHICH SATURATES	10.6G

COCA-COLA TẾT 2014
Rice Creative

Rice Creative is a multidisciplinary design agency founded in Saigon, Vietnam, in 2011 by Chi-An De Leo and Joshua Breidenbach. From huge global branding and advertising agencies, respectively, the duo sought a smaller, more personal environment to offer a new set of clients world-class creative solutions. The duo has now enlisted a small, handpicked team of talent to expand the reach of the agency. Rice has engaged with a number of like-minded companies both large and small to give them voice and visual identity. Rice plans to remain lean and focused. The approach is to form long-lasting bonds with select clients to produce powerful, effective, long-lasting creative work.

PRODUCT: Coca-Cola Tết 2014
CLIENT: Coca-Cola Co.
DESIGN FIRM: Rice Creative
ART DIRECTORS: Chi-An De Leo and Joshua Breidenbach
DESIGNERS: Chi-An De Leo, Joshua Breidenbach, Dinh Thi Thuy Truc, Huynh Tran Khanh Nguyen, and Nguyen Phu Hai
HAND LETTERERS: Joshua Breidenbach, Dinh Thi Thuy Truc, Huynh Tran Khanh Nguyen, and Nguyen Phu Hai
MEDIUM: Pencil, pen, and marker on paper
COUNTRY: Vietnam

PROJECT DESCRIPTION: Rice was asked to create a limited-edition can to help Coca-Cola celebrate the Tết holiday with their vast audience in Vietnam. Tết (Lunar New Year) is the most celebrated holiday in the country. It's a time for sharing good fortune, happiness, and New Year's wishes with family and friends. Rice aimed to create something meaningful, to touch people's hearts with a message.

Instead of one can, they created a series of Coca-Cola cans that spread millions of Tết wishes nationwide. They utilized Coca-Cola's Tết symbol, the swallow, as the device to give form to these wishes. For each can, hundreds of hand-drawn swallows were crafted and carefully arranged around a custom script, which together form a series of meaningful Vietnamese wishes:

"An" means peace.
"Tài" means success.
"Lộc" means prosperity.

These words are widely exchanged throughout the holiday and traditionally adorn Tết decorations. The simple act of sharing a Coke was transformed into a meaningful gesture. From supermarkets to small street-side shops, vendors took pride in decorating their shelves with Coca-Cola, and points-of-sale around the nation became festive environments. The brand loved that suddenly a Coke was a reminder and a carrier of the fundamental spirit of the holiday.

Can you talk about the significance of the swallows that you designed to work around the words?

CHI-AN DE LEO: The swallow is an ancient symbol widely used in Vietnam. Flocks congregate in the skies during spring, before the coming of the Lunar New Year. For that reason, they are iconic for the coming of the season. It wouldn't be spring if the swallows were not soaring. It is said that they "carry" the season in with them. This idea really resonated with us, as seen in the final design.

The custom script you created has a similar feel to the Coca-Cola logo. What were some of the other options that you played with in your explorations?

From the beginning we felt we wanted to emulate Spencerian script. But of course, we could not be sure without fairly vast exploration. We tried working with Coke's primary typeface, Gotham. It was a useful exercise, and at one point the recommendation from Coke's global office. In the end, comparing the script option to the sans-serif option, it was pretty clear which option "felt" better. The script flowed

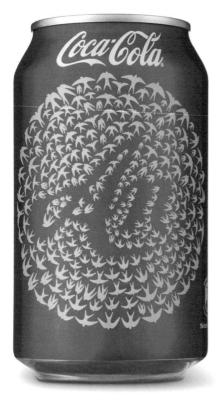
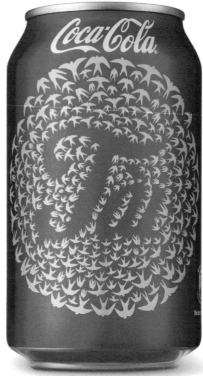

better. Coke had a concern that it looked too traditional, but eventually we all voted for the script.

What was the consumer response to your wonderfully festive packaging?

We had high expectations, knowing the potential of what we had created, and luckily we were not disappointed. We actually traveled to a couple different cites in Vietnam during Tết to see some of the millions of cans out in the real world. Small shops displayed the cans with pride and used them to create beautiful installations! The cans provided opportunity for billboarding, and from tiny street stalls to giant supermarkets, people had created their own advertisements with the cans. Some small, some huge! While the cans were out in the market, one of our photographer friends was sent to a very remote village in the far north of Vietnam. He sent us an image of the Coke can in a tiny stall, and it seriously moved us.

The words on the cans mean peace, success, and prosperity. Can you talk about why you chose those particular words for the three cans?

It was a real trick choosing the right words. In fact, there was a long list of possibilities, but the moment we noticed that those three words could be put together in any order to create an ever more powerful message, we knew we had finally nailed it. We had concerns that this might not even be possible, but we hoped that it would be.

Hand lettering on packages is a huge phenomenon in the West. Is the trend as strong in Vietnam?

Simply, no. This was a huge factor that kept us on pins and needles throughout the presentation process. The local taste would be for something much more slick. With an audience as big as Vietnam, Coke tends to keep their communication very universal. They aim for something they believe will be generally liked by millions of people from a huge range of demographics. We sought universality with personality. We were extremely surprised (shocked, actually) that the design ever saw the light of day.

Peace (An), Prosperity (Tài), and Success (Lộc) represent well wishes that are widely exchanged during the Tết holiday (Lunar New Year) in Vietnam. Each word is formed by the careful arrangement of hundreds of hand-drawn swallows.

97

The Tét packaging was a huge success in Vietnam, and was proudly displayed by shopkeepers across the country.

The swallow is an ancient symbol in Vietnam, representing the coming of spring. Early lettering exploration included sans serif as well as script options.

GARY TAXALI WOODEN TOY CLASSICS
Gary Taxali

G**T**ary Taxali maintains a studio in the Junction neighborhood of Toronto, Ontario, an area close to the junction of four railway lines called the West Toronto Diamond. The building itself is a converted 1929 felt factory. His studio has fantastic northern light, and to avoid distraction, Taxali has no telephone or Internet access. He does all his work there, from drawing and ideation to painting and screen printing final pieces.

TITLE: Gary Taxali Wooden Toy Classics
CLIENT: Indigo
HAND LETTERER/ILLUSTRATOR: Gary Taxali
MEDIUM: Mixed media
COUNTRY: Canada

PROJECT DESCRIPTION: Indigo, a large retailer in Canada known for selling books and gifts, approached Taxali and asked if he would like to reenvision classic toy designs in his own unique style. Taxali was given free reign of the project, but was asked to respect the design of the originals.

How did the idea for the toys develop?
GARY TAXALI: The idea for these toys is based on characters within my work. I used imagery from my fine art and from my other designer toy projects as a starting point. From there, I imagined how my characters would translate into the shape of the templates Indigo provided and made adjustments accordingly. I think the final results are interesting and exciting transformative objects based on the characters that people know from my work. Unlike previous toys I've created, these could actually be played with, so in that sense, they are true toys.

Who is the audience? Collectors, kids, etc.?
The audience for these toys includes small children and collectors alike. People buy them for an array of different purposes. Some choose to display them and preserve them as a collector's item, while others buy them as gifts for children.

What is your background as it relates to typography?
I employ as much typography as I can within my work. In this project, I incorporated my own font called Chumply in the package designs and also created handmade logotypes. Typography is an important aspect of my work. I treat it on the same level of visual and conceptual significance as I do drawn elements. In some cases, it's texture; other times, it's an important part of the idea. I love the interplay between words and pictures. To me, when the balance is right, it makes for wonderful harmony in a piece.

Where do you look for typographic inspiration?
My typographic inspiration comes from old packaging design, specifically from the 1930s and '40s. The type used in vintage posters and packages from across different cultures is exceptionally beautiful. I see inspiration everywhere, and sometimes it's in the most banal places, like a handmade sign for an auto body shop.

Who are some of the designers whose work you admire?
I admire the work of Chip Kidd, and also John Gall. Both designers make beautiful book jackets. They have classic design skills coupled with wit and sophistication that make for some of the most beautiful graphics I have ever seen. I have been lucky to work with both designers, who succeeded in bringing out the best in me.

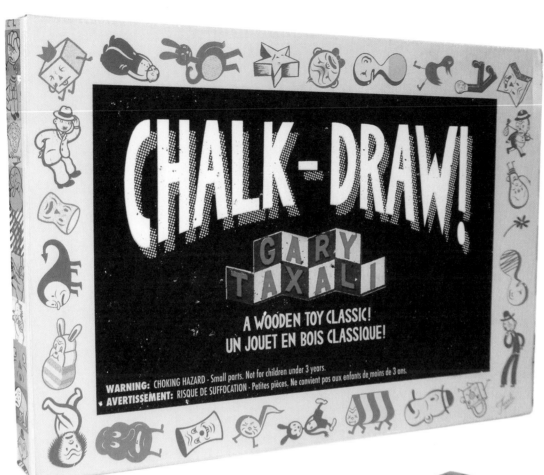

CHALK - DRAW!

GARY TAXALI

A WOODEN TOY CLASSIC!
UN JOUET EN BOIS CLASSIQUE!

WARNING: CHOKING HAZARD - Small parts. Not for children under 3 years.
AVERTISSEMENT: RISQUE DE SUFFOCATION - Petites pièces. Ne convient pas aux enfants de moins de 3 ans.

GARY TAXALI

A WOODEN TOY CLASSIC!
UN JOUET EN BOIS CLASSIQUE!

whirly!

*Taxali called this a "dream assignment,"
one that allowed him an incredible
showcase for his work, right down to the
package design, which he created himself.*

ABCDE FGHIJKL MNOPQRSTUVW XYZ!★

Whirly!

Ladder Fun!

Ladder-Fun

Taxali designed his own custom typeface, Chumply, and likes to use typography in both his commercial and fine art work.

PARTIZAN BREWING
Alec Doherty

Alec Doherty is a London-based illustrator working out of a studio located in an old warehouse originally used to make board games in the 1900s. Doherty studied graphic arts at Leeds Metropolitan University (now called Leeds Beckett University), and his clients include everyone from Microsoft to the Church of London. He collects cans of tomato puree and hopes to travel in space one day, when the tickets are cheaper.

PRODUCT: Partizan Brewing
CLIENT: Partizan Brewing
ART DIRECTOR/HAND LETTERER: Alec Doherty
PHOTOGRAPHER: Morgan K. Spencer
MEDIUM: Mixed media
COUNTRY: United Kingdom

PROJECT DESCRIPTION: Andy Smith, the owner and head brewer at Partizan Brewing, a small craft brewery based in London, asked Doherty to illustrate labels for his first few debut brews. Five labels turned into ten, and ten turned into 150, with the potential for more. The labels continue to evolve with Doherty's style, some elements becoming regular motifs that are replaced when everyone grows bored with them. It's a very free and open project that the illustrator loves.

Your illustrations tell a story. What was your thought process in arriving at your final solution?
ALEC DOHERTY: Partizan brews and bottles two or three new beers each week, which all have unique labels. Ideas for new labels come from different places. Sometimes it's related to the ingredients of the beer, sometimes the date it was brewed, sometimes a piece of music the brewers were listening to while they made it, or an event or person related to the brew. Other times it's something far more arbitrary, like a shape or range of colors that work well with the beer. The brewers let me know what they're brewing at the start of each week. I might have an idea about how I want the label to look straight away, but I occasionally need to take some time to think about it and let some ideas surface. I'll often do some research into the ingredients, dates, or style of beer, or I'll look through books and images to help me find some direction.

Your work involves lots of intertwined figures. How did this become such an integral part of your repertoire?
I think we're all naturally curious about each other, and I found myself drawing people a lot, almost exclusively—their faces, hands, and figures. I like how the body can be as expressive as the face. Dancing is a great example of that, so a lot of my characters are dancing.

How does the question of legibility come into play when working with hand lettering?
Legibility is something that comes up quite a lot with these beer labels. On a lot of them, the characters spell out the style of the beer with their bodies. It's not always easy to read the words. I like watching people studying the labels, figuring out the words and the little bits of symbology. For some people, figuring it out is rewarding, and they get a feeling of ownership for the time they've invested in that act; it's a bit of fun. I'm sure for a lot of people it's frustrating and they'd much prefer big, bold, legible type across the label, but that's not what I do.

Do you base your figures on a particular typeface?
I guess Futura features a lot in my work. I think it's a beautiful typeface with a very interesting history.

What was your favorite part of designing this project? It looks like such a fun one.
I really enjoy every aspect of the Partizan Brewing projects, from concept to artwork. I guess if I had to say in general, it would be that "golden hour" where the concept has come together and you have clarity, and all that's left to do is apply some polish to crisp everything up—I really enjoy that bit.

Doherty has illustrated more than 150 labels for London's Partizan Brewing.

"Getting ideas down on paper is important for me," Doherty says. *"It might just be half an idea that you revisit six months later with the other half. Ideas come from accidents when I'm sketching stuff out."*

Futura's clear, geometric shapes often provide the underlying framework for Doherty's human forms.

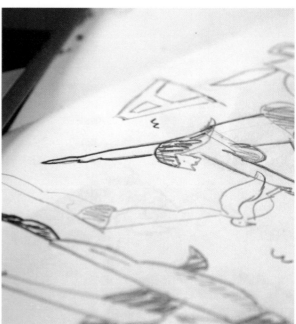

RIO COFFEE
Nate Williams

TITLE: Rio Coffee

CLIENT: Rio Coffee

AGENCY: Voice Design

ART DIRECTOR/DESIGNER:
Tom Crosby

HAND LETTERER/ILLUSTRATOR:
Nate Williams

MEDIUM: India ink, Sharpie,
and digital

COUNTRY: United States

Nate Williams enjoys making images, writing stories, and connecting with people. He recently returned to the mountains of Utah, when natal homing kicked in after a decade in Buenos Aires, Argentina.

PROJECT DESCRIPTION: Williams created a series of illustrations for Rio Coffee that represent the coffee bean's single origin.

You do a lot of lettering. How do you choose a particular style to work in?

NATE WILLIAMS: For the most part, I create lettering that is free-form, asymmetrical, organic, spontaneous, imperfect. If a client feels this tone is appropriate for them, they hire me and we talk about the specific creative direction for their project.

How do the lettering styles relate to the various countries that the coffee beans come from (e.g., Brazil, Kenya, etc.), if at all?

For Guatemala, I tried to capture the colonial influence. For Brazil, I was trying to capture the free-spirited Brazilian culture. For Kenya, I was referencing the textile and ceramic work. Finally, for Australia, the

type really wasn't specific to Australia. I was looking more for something that would integrate with the rest of the design.

Did you feel any need for this to correspond to the rest of the Rio Coffee brand?
No. I think the brand is defined by all of its parts, and not the other way around. It's like the saying, "actions speak louder than words." This specific packaging was intended to show the consumer where the coffee comes from in an authentic, visual way. All these small efforts (actions) will influence how consumers perceive the overarching brand.

How often do you use lettering reference (e.g., fonts, specimens, etc.), and where do you find them?
I don't really rely on reference. Of course, movie posters, cereal boxes, and all sorts of lettering have influenced me, but when I create type I usually just create spontaneously.

Williams's free-form, asymmetric letterforms were further imbued with cultural and artistic influences from each coffee's country of origin.

Single Origin Coffee
AUSTRALIA
BY RIO Coffee

BEANS
250g
NET

AUSTRALIA
Single Origin Coffee

A complex, dark-caramel
style coffee with marmalade
and chocolate hues, and
a dense, syrupy body.
Sourced exclusively from
a single plantation, this
coffee comes from the most
southern coffee growing
region in the world—
our very own backyard.

Since 1964 Rio Coffee has been brewing
Australia's passion for coffee, scouring
the world for the finest quality green
beans which our expert artisans roast
daily using traditional techniques.

riocoffee   View our full coffee portfolio
online at www.casario.com.au

Single Origin Coffee
GUATEMALA
by Rio Coffee

BEANS
250g
NET

GUATEMALA
Single Origin Coffee

Sweet, smooth and
well-balanced with a
creamy body and notes
of fruit and dark chocolate.
The fruit of the 'Maragogype'
coffee tree is often referred
to as the 'elephant bean'
due to its uniquely large
screen size. It is renowned
for its distinctive flavour
and mellow profile.

Since 1964 Rio Coffee has been brewing
Australia's passion for coffee, scouring
the world for the finest quality green
beans which our expert artisans roast
daily using traditional techniques.

riocoffee   View our full coffee portfolio
online at www.casario.com.au

FILIREA GI
Chris Zafeiriadis

A native of Thessaloniki, Greece, Chris Zafeiriadis studied graphic design at the Technological Educational Institute of Athens and is currently working as a graphic designer and illustrator at Bob Studio, also in Athens. He enjoys collaborating with other designers, illustrators, printers, and artists.

PRODUCT: Filirea Gi
CLIENT: Pashalis Zafeiriadis
DESIGNER/HAND LETTERER/
ILLUSTRATOR: Chris Zafeiriadis
MEDIUM: Silk screen and digital
COUNTRY: Greece

PROJECT DESCRIPTION: Zafeiriadis designed the packaging for his father's family wine. The fact that it is not sold, but rather given as a gift to friends and relatives, puts no restrictions on the design, leaving Zafeiriadis free to create whatever he wants each year.

How involved have you been in your family's winemaking tradition?
CHRIS ZAFEIRIADIS: My father has made wine every year since 1996. We have a small vineyard, and he is really passionate about it. He enjoys working year-round, taking care of it, plowing, pruning, etc.

The rest of the family and I help my father with the harvest and the grape pressing, which is done traditionally by pressing the grapes with our feet. After that, he finishes the winemaking process alone. We help him with the bottling after the wine has aged in wooden barrels for about a year.

How was the wine packaged before you created this?
In the early years, my father used to write just the harvest date on the bottle. As I grew older, I started making my first attempts at labeling it. Every year, I create a different package design. It is an opportunity for me to try different directions and design something with no limitations.

What has the response of the recipients been like?
Everybody in my family was very excited about the design. The fact that it was not a bottle with a classic label but had paper wrapped around the whole thing really made the difference. And of course when they saw all the winemaking procedures illustrated on it, they really fell in love.

Can you talk about the style of lettering that you chose?
I love illustrating. I did the lettering by hand, trying to match my illustrations. On most occasions, I start with the illustration and then I find the typography that best suits it.

The ultimate love-in: a family vineyard tended to by Dad, with contributions of grape pressing by Zafeiriadis family feet and package design by son Chris.

Φιλυρέα γη

ΚΤΗΜΑ ΠΑΣΧΑΛΗΣ ΖΑΦΕΙΡΙΑΔΗΣ

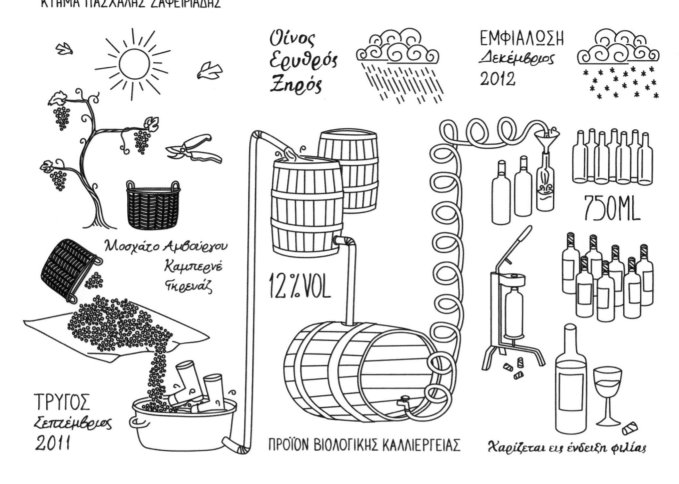

Οίνος Ερυθρός Ξηρός

ΕΜΦΙΑΛΩΣΗ Δεκέμβριος 2012

Μοσχάτο Αμβούργου Καμπερνέ Γκρενάζ

12% VOL

750ML

ΤΡΥΓΟΣ Σεπτέμβριος 2011

ΠΡΟΪΟΝ ΒΙΟΛΟΓΙΚΗΣ ΚΑΛΛΙΕΡΓΕΙΑΣ

Χαρίζεται εις ένδειξη φιλίας

Zafeiriadis silk-screens each bottle by hand, and recipients delight in seeing the illustrated winemaking process on the paper wrapper.

Προϊόν Βιολογικής
Καλλιέργειας

Τρύγος
Σεπτέμβριος
2011

Καμπερνέ
Γκρενάς
Μοσχάτο

12% vol

Εμφιάλωση
Δεκέμβριος
2012

Οίνος
Ερυθρός
Ξηρός

750 ml

Φιλιγκέα γη

Κτήμα Πασχάλης
Σαφειριάδης

Χαρίζεται
εις ένδειξη φιλίας

YUME UMĒ
Kelly Dorsey

A branding and creative services agency with offices in Philadelphia, Newport Beach, CA, and Gainesville, FL, 1600ver90 has vast experience generating results for nationally recognized brand leaders. 1600ver90's client roster spans consumer brands, among them Nike, Godiva, Under Armour, American Eagle Outfitters, and Ferrari North America; professional sporting teams including the Philadelphia Eagles, Miami Dolphins, and Los Angeles Angels; and an esteemed roster of higher education institutions, including UCLA, Duke University, University of Notre Dame, and the University of Florida. Services range from the development of single creative pieces (print collateral, packaging, and websites) to institution-wide image and rebranding campaigns. In 2013, AgencyPost.com named 1600ver90 to its "Agency 100" list of fastest-growing agencies in America.

PRODUCT: Yume Umē
CLIENT: Yume Umē
DESIGN FIRM: 1600ver90
CREATIVE DIRECTOR:
Steve Penning
DESIGNER/HAND LETTERER:
Kelly Dorsey
ILLUSTRATOR: Mike Gray
COPYWRITER: Kyle Arango
PHOTOGRAPHER: Tim Beitz
MEDIUM: Pen and ink
COUNTRY: United States

PROJECT DESCRIPTION: Based out of Gainesville, FL, Rolls n' Bowls was a made-to-order Japanese kitchen. The restaurant was looking to make the leap from local stronghold to national franchise. 1600ver90 was tasked with developing a brand and identity to help them do so, and created Yume Umē.

Can you talk a bit about the copywriting for Yume Umē? The puns are genius.
KELLY DORSEY: The concept we created was born out of the idea that you can create whatever you dream up. We thought about Yume Umē as a world filled with fun characters that all have their own personalities. The puns are our take on how they speak to the outside world and are intended to put a smile on the face of restaurant goers. Honestly, it was really effortless collaborating with the copywriter; we were just trying to have fun.

What has the customer response been to the new branding?
I've heard from the creative director that "it has been positive." Not sure what that means! I left the agency before this work was rolled out at the new franchises. As far as I know they are still a client, and 1600ver90 has since opened an office in Gainesville.

Did the client actively seek out hand-drawn typography?
They did not actively seek out hand-drawn typography. The hand-drawn aspect was largely driven by our concept. "Yume Umē" is Japanese for "Yummy Dream," and it emphasizes the role of imagination and creativity in the dining experience customers have at the restaurant, where "Dreams Are

Edible." We brought this to life with an organic illustrated approach. It was important to me that the design also have some sort of tie-in or nod to the tradition of artmaking in Japan. This is how the execution of the logo came about. It's meant to feel like an artist's mark or seal, which usually appears as a red stamp of sorts on the bottom of an artwork. I wanted the logo to feel like that artist's mark when applied to the overall branding.

Do you tend to do a lot of notebook sketches on all your projects, as you have for Yume Umē?
To be honest, I haven't done much sketching like this for a couple of years. Yes, I always have a sketchbook, but as my career has turned more toward the strategic side of making, my sketchbook has slowly morphed into more of a notebook, containing notes, client feedback, directives, and thoughts—just words written hastily. I still sketch out visual executions here and there, but not like I used to. I haven't done any hand-drawn lettering in so long that when I sat myself down to do some save-the-dates for my upcoming wedding, it really did seem like the skill was far away from me. There was definitely some procrastination due to the anxiety of being expected to produce the work I am known for, and realizing that I've really not kept up on my craft. It's kind of like riding a bike, but I had to think about it more than I used to.

120

121

Dorsey kept the spirit light throughout dozens of
pages of character and logo explorations.

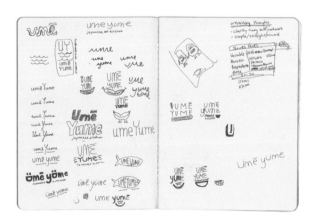

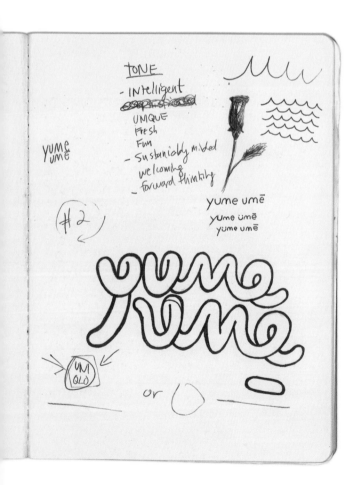

MIC'S CHILLI
Steve Simpson

For thirty years Steve Simpson has been applying his multidisciplinary skills to creative projects for a diverse range of clients from across the globe. His innovative, award-winning approach to graphic design, typography, and illustration is built on fresh thinking, traditional skills, and a dose of fun. Simpson lives on the east coast of Ireland, where a good sense of humor is essential.

PRODUCT: Mic's Chilli
CLIENT: Mic's Chilli
ART DIRECTOR/DESIGNER/HAND LETTERER: Steve Simpson
MEDIUM: Pencil and digital
COUNTRY: Republic of Ireland

PROJECT DESCRIPTION: Michael Wejchert, the creator of Mic's Chilli, a hot sauce brand, approached Simpson in the early stages of product development in order to discuss some ideas for design. Wejchert had quite an open brief and granted Simpson a good deal of control over the process. Wejchert, though, had two requirements: that the labels strongly feature characters, and that the label be rich enough to be a talking point. He hoped that the bottles would be featured on dining tables as a discussion piece rather than living in the cupboard. Other than that, Simpson was given complete freedom to design both the label and the Mic's Chilli logo.

At what stage do you start developing a specific letter style? Do you have styles in mind at the beginning? Or is it strictly compositional at that stage?

STEVE SIMPSON: To my way of working, illustration and lettering style are intrinsically linked. I tend to develop letter styles at the outset, however roughly. It is a balancing process, trying to create a combination of letter styles that are pleasing to the eye as well as complementary to the design.

I approached this label as if I was creating one large illustration rather than a design with an embedded illustration, as is more often the case. My vision was to have the whole label illustrated and to use hand lettering as much as possible. This led to illustrating the barcodes as well.

Your playful, cartoony lettering evokes vintage type (and artistic, like Posada). Can you talk about your influences, and how you've channeled those sources?

I have always had an interest in classical lettering. My boyhood home, in the north of England, is built on Roman burial grounds, and I spent lots of time with my dad digging up remnants. I used to pour over these finds, and this led to an interest in Roman and early Christian lettering and stone carvings of the period. I tended to pick up letterforms in an organic way over the years and automatically stored them in my mind to be drawn on when needed. I never feel as if I am shoehorning them into place.

Also, I studied graphics at college at a time before desktop publishing, which involved a lot of hand lettering, and continue my study of typefaces to this day.

How directly do you reference those lettering sources? Is it more an osmosis thing? Or like many lettering artists, do you build from existing classic forms?

I usually draw on my bank of mental imagery when referencing lettering. Occasionally, I come across a particular letter that does not work out, and at this point I reference a classical source, like William Morris's letterforms or early Christian carvings. More often, though, I'll be visiting a museum and come across some inspirational lettering that leads to me getting my sketchbook out; when a suitable project comes up, it will find its way into my work.

OPPOSITE: *Simpson is "hugely influenced by a mixture of classical and medieval letterforms, often combined with vintage 1950s advertising and a dash of William Morris."*

The chisel serif logo appears across the entire product line, evoking the spirit of the American Southwest and wanted posters.

INFERNO
SAUCE

 lite

Before desktop publishing made it possible for the most casual user to create graphic designs—before the instant gratification of twenty-four-hour copy shops—a tremendous amount of graphic work was generated by jobbing print shops. These shops considered the graphic arts to be a blue-collar trade in which journeyman printers used a standardized and reliable set of tools to produce sturdy, functional design. The pieces that follow channel that utilitarian aesthetic, while simultaneously calling up the spirit of DIY, fearlessly fusing white-collar design with blue-collar craft.

THE HIDDEN SEA
Jon Contino

Jon Contino's work combines Old and New World aesthetics into a modern, minimalist style. It reveals the influences of historical New York, contemporary street art, and the lost art of hand-drawn lettering. Through these images, he tells his story as a designer and consultant to brands ranging from the smallest boutique start-ups to the largest corporations.

PRODUCT: The Hidden Sea
CLIENT: Co Partnership
DESIGN FIRM: Jon Contino Studio
CREATIVE DIRECTOR: Max Harkness
DESIGNER/HAND LETTERER/ILLUSTRATOR: Jon Contino
MEDIUM: Ink
COUNTRY: United States

PROJECT DESCRIPTION: Co Partnership approached Contino to create an identity for The Hidden Sea, a South Australian wine brand. He was asked to expand on an illustration that he originally designed for his menswear brand CXXVI Clothing Co. The particular piece they referenced was one Contino thought could've been done much better, and he was excited to be able to go back to the beginning and reimagine the initial idea. Once the style was locked down, Contino designed the remaining identity and packaging systems fairly easily. Since the language was so clearly defined, the rest fell into place. "There's a lot of natural, ecological, Australian history behind the brand," Contino says, "so it was fun to explore that using design."

Can you talk about your lettering process for The Hidden Sea?
JON CONTINO: I really wanted to make it clear that this brand had a deep connection to the ocean. My solution was to incorporate wavelike flourishes into the decorative elements of the type. I tried a few different styles, but in the end, the whole thing had to be pretty tightly composed so that the waves didn't distract from the legibility of the words themselves.

Tell me about the monogram. Did you consider the ubiquitous anchor for the "T"?
Absolutely not. I think the anchor iconography has been done to death. I'm absolutely guilty of it as well, but after the however-many-hundredth client asked me to do something with an anchor, I had to abandon it forever. I think the last anchor I ever incorporated into a design was in 2010. I just had to put a stop to the madness. To this day, four years later, I'm still seeing people saturate the market with anchors. It's a beautiful shape, but it's basically lost its meaning.

A nautical theme is tough to pull off, since people just assume anchors all the time, so I try to go for the more obscure elements that might not specifically speak to the ocean but create an environment of unmistakable intentions. Thanks to the whale, though, this design had a little help, so that can't be ignored either.

How far did the client want to take the nautical theme (e.g., the rope box handles)?
The client wanted to make sure the nautical theme was clear, but that didn't mean it had to be super graphic. One of the great things about designing an entire package is that you can use alternate means for describing a theme. The rope on the box is one of those things that isn't slap-you-in-the-face nautical but, when used properly, becomes something very obvious.

What were some of your typographic influences on this project?
I combined the bones of classic Spencerian calligraphy with the soul of traditional tattoo-style script. Mixed in with a little bit of those wave flourishes and it became a beast in its own right.

The concept behind The Hidden Sea is loosely based on an illustration originally created for Contino's menswear brand, CXXVI Clothing Co. Expanding on the original artwork, the illustration matured into a complete brand featuring a full identity and packaging system.

How much direction came from the client? Is there a story behind "The Limestone Coast"? How did that shape the direction?

The identity is actually based on the Limestone Coast of Australia, which has become quite famous for its ancient whale fossils. That's a pretty unique thing to find, so the client really wanted to use the whale as the foundation for the brand. The fossils are basically part of the earth from which the grapes grow to create the wine.

You have a distinctive style. Did the client seek you out for your rustic approach?

Absolutely. It was clear from the beginning that my style was the one thing that was a defined parameter of the project. The rest was up for grabs as we explored and experimented with ideas and execution.

"The labels were a fun challenge, since I had to reimagine the script within the allotted space over and over again," says Contino. "It was a pretty unique process trying to maintain the vibe while keeping it clear that it's a different variety."

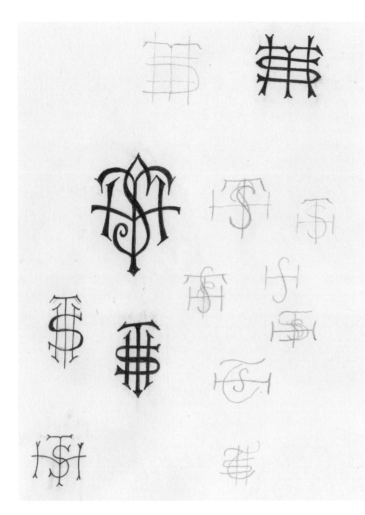

Viognier 2012 → Riesling 2012

Semillon Sauvignon Blanc

Viognier

Riesling

Merlot 2010   Sauvignon Blanc 2011

Chardonnay 2011

Rosé 2012   Shiraz 2010

Cabernet Sauvignon

Rosé

Chardonnay

135

R&B
Saint Bernadine Mission Communications Inc.

S aint Bernadine Mission Communications Inc. is a Vancouver-based agency specializing in branding, advertising, graphic design, and digital design. Saint Bernadine is the patron saint of advertising and graphic design.

PRODUCT: R&B
CLIENT: R&B Brewing Co.
DESIGN FIRM: Saint Bernadine Mission Communications Inc.
CREATIVE DIRECTORS: Andrew Samuel and David Walker
ART DIRECTORS/DESIGNERS/HAND LETTERERS: Rory O'Sullivan and Helen Eady
MEDIUM: Ink and digital
COUNTRY: Canada

PROJECT DESCRIPTION: R&B Brewing is one of Vancouver's original East Van microbreweries, predating the current trend by almost two decades. Partners Rick ("R") and Barry ("B") asked Saint Bernadine to help with a packaging redesign, necessitated by the move from 650mL bottles to a standard 341mL six-pack format.

The design language is inspired by R&B's resolutely handcrafted product—as such, every element on the packaging is rendered by hand, including the bar code. Bold silhouette key illustrations take their cues from the beers' quirky naming conventions, and strong colors aid standout and shelf blocking. Key brand story elements, hand rendered in expressive type, and additional information like IBU scale, boil information, aging temperature, hop and barley type, and mix all support R&B's independent craft positioning.

What does hand lettering bring to your work that standard or initially hand-lettered fonts could not?
ANDREW SAMUEL: Hand lettering brings an energy and personality to the content. It reflects the handcrafted nature of R&B's product as well as their enthusiasm. The hand lettering marries the typography and the illustration, creating a cohesive unit. From a practical standpoint, hand lettering allows for maximum flexibility of content.

How much is the hand lettering closely based on direct reference to existing typefaces?
Some of the hand-drawn type is based on actual fonts, some are inspired by actual fonts and modified accordingly, and many are custom. Traditionally inspired letterforms bring an element of structure and legibility to the necessary content, while custom lettering brings an illustrative quality.

R&B Brewing's packages have a lot of text. What challenges did you face with editing and making corrections since all of the type is hand done, including the small print?
As with any project, there were changes to content during the design process. There was no easy way to avoid the fact that we had to redraw type on a couple of occasions—a labor of love.

To what do you attribute the popularity of hand lettering on packaging?
Hand lettering brings warmth and personality to a product and its content—it removes a layer of formality. Hand lettering engages the consumer and makes the brand feel approachable and familiar. So, for the right project, it's an invaluable option.

Every element on the final packages was created by hand, all in support of R&B's independent craft positioning. It's easy to see the love that went into R&B beers.

R&B Brewing calls itself "Vancouver's local microbrewery," and has made handcrafted ales and lagers in the heart of East Vancouver's historic Brewery Creek since 1997.

Rick & Barry INVITE YOU TO TRY OUR AWARD-WINNING EAST SIDE BITTER WHICH IS CAREFULLY HANDCRAFTED IN SMALL BATCHES FROM OUR VANCOUVER MICROBREWERY. R&B BREWING COMPANY, VANCOUVER, BC. EAST VANCOUVER'S ORIGINAL LOCAL BREWERY.

CAREFULLY Handcrafted IN SMALL BATCHES FROM OUR **EAST** VAN MICRO BREWERY | R&B BREWING COMPANY EAST VANCOUVER'S ORIGINAL LOCAL BREWERY

WE'VE BEEN *brewing* FINELY MADE HAND CRAFTED ALES & LAGERS IN THE HEART OF EAST VANCOUVER'S HISTORIC Brewery Creek district for over a decade with a commitment to keeping things MODEST, INDEPENDENT & UNIQUE

EXTRA SPECIAL BITTER

MEDIUM 13 srm

90
9

90 minutes

2 weeks at 5°C

Citra, Cascade, Columbus
Light Crystal, Cara-Pils, Munich

27 5.5% CANADIAN BREWING AWARDS

Ridiculously HOPPED

DID YOU KNOW?
You can get draught beer for your party.
~ r-and-b.com ~

Think Think

Drink Drink

**THINK INDEPENDENTLY
DRINK INDEPENDENTLY**
341ml | 5%alc/vol | beer/bière

Detailed sketches document the creation of the beer packaging.
"From a practical standpoint," says Samuel, "hand lettering
allows for maximum flexibility of content."

NAGGING DOUBT
Tanamachi Studio

Tanamachi Studio is a Brooklyn-based boutique graphic design studio specializing in hand lettering and custom typography for editorial, lifestyle, food, and fashion brands.

PRODUCT: Nagging Doubt
CLIENT: Nagging Doubt
AGENCY: Brandever
DESIGN FIRM: Tanamachi Studio
ART DIRECTORS: Bernie Hadley-Beauregard and Laurie Millotte
DESIGNER/HAND LETTERER: Dana Tanamachi-Williams
PHOTOGRAPHER: Laurie Millotte
MEDIUM: Chalk
COUNTRY: United States

PROJECT DESCRIPTION: Tanamachi Studio was asked to design wine labels for Nagging Doubt, which is based out of Okanagan Falls in British Columbia.

You're known for chalk lettering. Do you always work in that medium, even when the end product is digital?

DANA TANAMACHI-WILLIAMS: Truthfully, I rarely work in chalk these days. It's probably been at least six months since my last installation. I'm currently focusing on and enjoying other mediums: paint, pen, pencil, etc. All still done by hand, of course! If the end product is digital, I typically scan the artwork in and convert it to vector, though I usually push to keep the artwork as an image or photograph to keep the textural qualities.

How do you keep it fresh and distinguish yourself from the many imitators?

Imitators can be a great source of inspiration! Instead of complaining, I channel that energy into new projects, experiments, and mediums. I'm so thankful my clients are always along for the ride. The best thing about creating original work is that you have a unique story that no one else has. Chalk started out as a personal hobby and turned into something much more. And that's how I constantly look at each new season of my career—it's the inspired personal work that's going to keep you moving, evolving, and one step ahead.

How do you find a balance between crisp graphic readability and maintaining the hazy chalk effect (especially in print)?

For these Nagging Doubt labels, I did them at such a large scale so I could get the kind of details that would greatly enhance readability once photographed. If I would've tried to create these at actual wine label size, there's no way I could have gotten even 90 percent of the detail that's in the final.

Does the chalk process allow the sketch to become the final art (i.e., do you draw, erase, draw, erase, gradually transforming the sketch into a final)?

Yes, exactly. If I'm working in chalk, the initial sketches I send to clients are usually terrible. But, I'll send them photos along the way as I develop the actual art on a board or wall. It's fun to hear the relief in their voices once they see the final art come together. I appreciate that they trust me!

What are the challenges of being so known for a particular style?

I'm not sure there are too many challenges, honestly. Literally every single one of my clients has been open to growing with me. I'm a designer first and foremost, and a few years ago, my lettering just happened to take the medium of chalk. I'm definitely not a chalk artist; as we all know that's something completely different. You're never going to see me creating 3D chalk drawings on a sidewalk! I'm a designer. Designers typically have a certain set of skills that can be applied to any medium. Give any letterer a paintbrush, crayon, or piece of charcoal and I guarantee they'll make something beautiful out of it. I would say that my personal style may be fairly recognizable to some designers, but I bet they could recognize it without it being in chalk.

OPPOSITE: *"You'll notice just how much information is on these labels," Tanamachi-Williams says. "It was a challenge to include all the descriptors that the client wanted while still making the labels beautiful and legible."*

The labels were created on 4' x 7' chalkboards, then photographed and reduced to size. The QR (Quick Response) code in the corner of the label links to a time-lapse video of the work being created.

The evolution of rough sketches to finished designs highlights how clear Tanamachi-Williams is in her thought process. Initial drawings are tightened up and refined, but not redesigned.

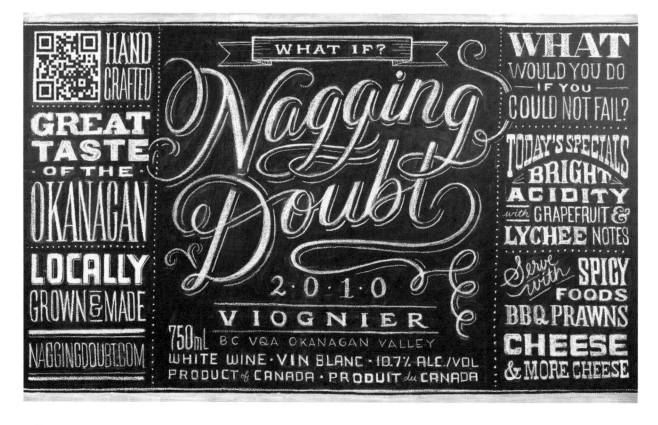

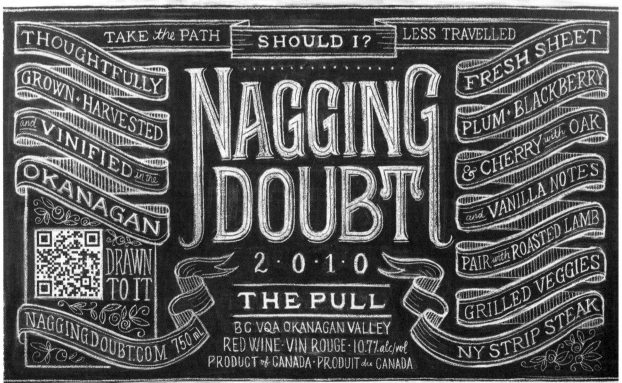

MIAMI COCKTAIL CO. SMALL BATCH ORGANICS

Mary Kate McDevitt

Mary Kate McDevitt is an illustrator and hand letterer based in Brooklyn. She approaches lettering with a vintage style and a playful twist. Clients include Nike, Sesame Street, Puffin Books, Adobe, and Chronicle Books.

PRODUCT: Miami Cocktail Co. Small Batch Organics
CLIENT: Miami Cocktail Co.
CREATIVE DIRECTOR: Omar Morson
DESIGNER/HAND LETTERER: Mary Kate McDevitt
MEDIUM: Pen and ink
COUNTRY: United States

PROJECT DESCRIPTION: McDevitt worked with Miami Cocktail Co. to create hand-lettered labels for their new cocktails. The cocktails are small batch with a focus on organic, natural ingredients, so they wanted to reflect those qualities in the design.

Were you involved in creating the "M" logo embossed in the glass at the top of the bottle?
MARY KATE MCDEVITT: I didn't create the logo, but I did redraw it so it had a similar hand-drawn edge as the rest of the label.

Was it a challenge to fit words of different lengths (e.g, "Mojito," "Margarita") into a fixed width?
"Margarita" was the biggest challenge to fit; all the other flavors fit so perfectly. For a round or two of revisions, I had "Margarita" at a shorter height because I was concerned with the legibility. In the end, we decided to make it all the same height so everything is cohesive (plus, it looks much better together on the shelves this way).

There is a lot of ingredient lettering. Do you create whole alphabets in advance? Or did you just make it up as you went?
There was a lot of attention paid to the ingredients because the client really wanted the natural and organic ones to be featured rather than set in small type and at the bottom. The lettering for the ingredients was all custom because we wanted to call out certain words in different styles and insert some illustrations of fruit or ornamentation.

How do you address issues of scale/weight/ legibility between small and large letters?

Since there was such a difference between the lengths of the words, like "Margarita," I couldn't use an ornate style; it would have been so hard to read. I still wanted the labels to have a special look, so I went with a rounded serif and relied on the lock-up to carry that unique quality.

Do you draw at size? Or bigger, then reduce?
For these labels, I drew them bigger then reduced them. Most of the time I draw at size. The issue that comes up when you draw bigger and then reduce is making sure the line weights will translate when the scale shifts. So I just had to review the drawings at the correct scale.

The final is more casual and less rustic than some of your sketches. Talk about those variations and the evolution of the approach.
The brief was to make the brand look specialty, organic, and hand drawn. I did some variations on simple hand-drawn type with special accents, and some with a focus on uniqueness. I loved the script and ornamental and illustrative sketches, but the client ultimately felt that the script and more ornamental sketches were too fancy. In the end, I was very pleased with the look and feel of the label, which fit with the brand and concept.

The lettering was drawn larger, then reduced, with careful attention paid to line weights so proper proportion and legibility were maintained. The logo was also redrawn by hand to integrate more gracefully with the rest of the package.

Miami
Cocktail Co.

Miami
Cocktail Co.

Miami
Cocktail Co.

ORGANIC
100% NATURAL
ARTISINAL

Made in USA
750ml | 9% alc. vol

ORGANIC PREMIUM RED WINE
LIGHTLY INFUSED WITH TROPICAL
FRUITS of PINEAPPLE ORANGE
LIME, KIWI, PAPAYA, and MANGO

ORGANIC
100% NATURAL
ARTISINAL

Made in USA
750ml | 9% alc. vol

ORGANIC PREMIUM RED WINE
LIGHTLY INFUSED WITH TROPICAL
FRUITS of PINEAPPLE ORANGE
LIME, KIWI, PAPAYA, and MANGO

ORGANIC
100% NATURAL
ARTISINAL

Made in USA
750ml | 9% alc. vol

ORGANIC PREMIUM RED WINE
LIGHTLY INFUSED WITH TROPICAL
FRUITS of PINEAPPLE ORANGE
LIME, KIWI, PAPAYA, and MANGO

The more embellished lettering and ornamentation of these early sketches was eventually discarded in favor of a less "fancy" aesthetic.

Miami Cocktail Co.

Small BATCH

SANGRIA

·ORGANIC·
100% Natural
·ARTISINAL·

ORGANIC
·100% NATURAL·
ARTISINAL

=Made in USA=
750mL · 9% alc.vol
Organic Premium RED
Wine lightly infused WITH
TROPICAL FRUITS of PINEAPPLE,
ORANGE, lime KIWI, Papaya
– and MANGO –

Miami Cocktail Co.

Small Batch

SANGRIA

ORGANIC
Hand-Crafted
100% NATURAL
ARTISINAL

·ORGANIC·
100% NATURAL
·ARTISINAL·

=Made in USA=
750mL · 9% alc.vol
Organic Premium RED
Wine lightly infused WITH
TROPICAL FRUITS of PINEAPPLE,
ORANGE, lime KIWI, Papaya
– and MANGO –

Miami Cocktail Co.

SMALL BATCH

SANGRIA

ORGANIC · HAND-CRAFTED
100% NATURAL · ARTISINAL

·ORGANIC·
100% NATURAL
·ARTISINAL·

Made in USA
750mL | 9% alc.vol

ORGANIC PREMIUM RED WINE
LIGHTLY INFUSED WITH TROPICAL
FRUITS of PINEAPPLE ORANGE
LIME, KIWI, PAPAYA, and MANGO

153

McDevitt's pencil sketches are tightly rendered, though her style is informal and filled with personality.

CHARLIE'S QUENCHER
Brother Design

Brother Design specializes in private label, brand, and packaging design and is based in Auckland, New Zealand. Brother won the 2014 Vertex Awards Best of Show and four gold medals for Pams Private Label, where they outshone designs for global retail giants such as Tesco, Safeway, and Woolworths.

PRODUCT: Charlie's Quencher
CLIENT: Better Drinks Co.
DESIGN FIRM: Brother Design
ART DIRECTOR/DESIGNER:
Paula Bunny
HAND LETTERER/ILLUSTRATOR:
Chris Piascik
MEDIUM: Pencil and digital
COUNTRY: New Zealand

PROJECT DESCRIPTION: Charlie's was launched in 1999 by three New Zealanders. They started out by juicing oranges by hand and sampling the juice in supermarkets. It's since become an iconic New Zealand brand with six different product ranges, sold in both New Zealand and Australia.

Brother was asked to design a fresh new label that redefined and rejuvenated the existing Charlie's Quencher range, as well as to help rebuild the Charlie's brand personality and improve shelf impact. Charlie's had always done things a bit differently than their competitors, with their brand proposition being "honest goodness."

Brother's main objective was to communicate that the juice is "un-messed with" (i.e., less manufactured, more pure, more real) while maintaining a bit of the "larrikin" (young hoodlum), which is a core part of the Charlie's DNA.

Hand-drawn type is part of the design zeitgeist in the United States. Can you talk about its popularity in New Zealand?
PAULA BUNNY: Hand-drawn type is very popular in New Zealand, and has been for what seems like forever. We have a pretty laid-back kind of lifestyle, so we tend to not get too fussy or too formal; we like things to be a bit simpler, and not too complicated. We relate to the freeness and honesty of hand-drawn type. It's unpretentious—just like we like to think we are!

Handcrafted type is a great way to humanize a brand and give it a handmade sensibility. It moves the product away from feeling manufactured or processed or too slick and overdone. It's a graphic resource we've used for many years on the private label brand Pams, which we design.

Who is the typical consumer, and why would the style of writing you chose appeal to them?

The typical Charlie's consumer wants an uncomplicated, healthy, honest juice experience…with some attitude. They look for a brand that cuts through the clutter of the refrigerator case and gives them the facts straight up, but in a cheeky, good-hearted, spirited way—therefore a loose, relaxed handwritten style and tone was appropriate. It helped to reinforce our good-natured, uncomplicated personality— where we New Zealanders like to tell it like it is and not take ourselves too seriously.

Your sketches include a guy on a motorcycle. How did the packaging evolve to not include this figure on the front of the juice bottle?
The guy on the motorcycle was on the very first Charlie's label that came out in 1999, when the company was just three guys who were juicing oranges inside supermarkets. I remembered this little guy, as I had just arrived back from eight years living in New

York and the label stood out to me as being so fresh and relaxed and casual—unlike anything I'd seen in the United States prior to that time. I thought it would be great to bring this fella back, as he symbolized the casual roots of this iconic New Zealand brand, and I did look at including him on some of my earlier sketches, but it was decided that we would focus instead on communicating the old-fashioned lemonade style of the product on the front label. Motorcycle guy is on the back label, though, as a fun little reminder of where Charlie's had come from—three guys racing 'round to supermarkets in Auckland, New Zealand, on a scooter with their fresh squeezed orange juice.

What was your inspiration for the hand-drawn typography? Historical references?
They've always fronted the Charlie's ad campaigns with Mark Ellis (a New Zealand personality/ex–All Blacks rugby player), who is one of the original owners. They have always used his distinctive language in these ads and on packaging as well.

The language and the look are kind of

New Zealand back in 1985 (Redline bikes, high-top sneakers, Walkman cassette decks, etc.)—very lo-fi.

I started researching illustration styles and hired Chris Piascik, an illustrator based in the United States whose style felt so right: unfussy, not too slick, and a little crazy/quirky. Charlie's has always been lo-fi—a bit raw and rough around the edges, unpretentious—and this has always been a really good way to communicate the "un-messed with," natural, unprocessed ethos of Charlie's.

Who came up with the "Yep" line?
I wanted to create a kind of a lighthearted, friendly conversation to explain the style of the product, which is an old-fashioned lemonade style, and I wanted that tone to be similar to Mark Ellis's distinctive kiwi blokey voice. So that's how "yep" came about—it's exactly how Mark would have said it; he would have said "Yep…it's an old-fashioned Charlie's lemonade Quencher." It's a warm, friendly way of inviting the consumer in. It's how we kiwis talk.

The cheeky and good-hearted personality of this type expresses the casual roots of this New Zealand juice brand.

158

Brother Design's various directions position Charlie's as classic
and old fashioned, playing off the product's DIY roots.

FETZER CRIMSON AND QUARTZ
CF Napa Brand Design

CF Napa Brand Design specializes in brand strategy and design exclusively for the wine, spirits, and beer industries. With more than thirty-five years of experience, their capabilities include brand strategy, naming, logo development, package design, marketing and point-of-sale collateral, signage, and websites.

CF Napa has created hundreds of new brands with incredible success and restaged even more existing brands, maximizing their strategic positioning. Their clients span the full gamut from growers and producers to retailers.

PRODUCT: Fetzer Crimson and Quartz
CLIENT: Fetzer
DESIGN FIRM: CF Napa Brand Design
CREATIVE DIRECTOR: David Schuemann
DESIGN DIRECTOR: Kevin Reeves
DESIGNER/HAND LETTERER/ILLUSTRATOR: Dana Deininger
PHOTOGRAPHER: Tucker & Hossler Photography
MEDIUM: Linocut, pencil, and ink
COUNTRY: United States

PROJECT DESCRIPTION: CF Napa extended the Fetzer brand with a new pair of Fetzer red and white wine blends, Crimson and Quartz. These celebrate the Fetzer brand's natural, carefree—some might say "hippie"—personality. In line with Fetzer's boutique, handcrafted spirit, the agency hand drew the entire label, including all the type. The style of the labels simulates the layered look of silk-screening, accentuated even further by multiple layers of embossing. Crimson and Quartz now stand out beautifully on the shelf and have proven so popular that the brand has grown into a full line of wines.

Did your design originally have the Fetzer logo (bottom left in final design), or was the client concerned that this new approach was too far removed from their core brand, and asked for the logo?

DAVID SCHUEMANN: We explored designs both with and without the Fetzer logo. Ultimately we agreed with the client that connecting the new sub-brand to Fetzer would provide the desired halo effect for Fetzer with this new millennial-focused wine.

Some earlier sketches suggest a more elegant, Victorian lettering approach. Why was this rejected?

Our focus became creating a design expression that would attract millennial wine drinkers and still appeal to baby boomers through a 1970s feel.

Did the client request hand lettering, or were other options considered?

The hand lettering was our idea. We presented eight to ten concepts, but this one just felt right. It struck the right balance between a fresh, handcrafted sensibility and a throwback '70s vibe.

Hand lettering has been a design staple for the last several years. Where do you see what was once thought to be a trend heading in the future?

I guess we never thought of it as a trend. Graphic design has its roots in hand-drawn type and design—in our studio we draw everything before we go to the computer. Sometimes, like on this project, the hand-drawn versions are better or more appropriate than anything we could "clean up" on the computer. I don't see the human desire for connection with others disappearing, and written language/type is an expression of this desire. The digital age has only amplified this need, and I suppose that's why hand calligraphy has become so popular in recent years around the world.

OPPOSITE: *A modern but slightly retro type style was created by simulating the richly layered look of silk screen.*

Some of the preliminary sketches reflect a more traditional, artisanal approach to wine label design.

DENNIS MARTIN
Dennis Martin
M M M M

WINE-
MAKER

Wine-
Maker

Dennis Martin
wine
Maker

DENNIS MARTIN
DENNIS MARTIN
M M M M

DENNIS

PINOT NOIR
PINOT NOIR

DENNIS MARTIN
wine
Wine
maker   Maker

FETZER
CRIMSON

PROPRIETARY

PROPRIETARY
RED BLEND *
PROPRIETARY

WINEMAKER'S FAVORITE

RED BLEND

2009 VINTAGE Aged **14** months in OAK barrels

7 750 750 ML. ML.

NORTH COAST O

13.5% ALC. BY VOL.

=
CA
=
hand Crafted

VARIETAL CABERNET SAUVIGNON

13.5% ALC. BY VOL.

MADE —IN—

MENDOCINO COUNTY

The final direction chosen is more contemporary and youthful than CF Napa's initial sketches, making it perfect for millennials as well as the typical baby boomer buyer.

BIENBEBIDO
Moruba

Moruba was founded by Spanish designers Daniel Morales and Javier Euba. Their passion for design truly blossoms when they are able to work with the complete confidence of their clients, often elevating the work beyond the initial objectives. Moruba has been recognized by national and international awards, and published in books and magazines.

PRODUCT: Bienbebido
CLIENT: Vintae
DESIGN FIRM: Moruba
CREATIVE DIRECTORS/
DESIGNERS/HAND LETTERERS/
ILLUSTRATORS: Daniel Morales
and Javier Euba
MEDIUM: Silk screen
COUNTRY: Spain

PROJECT DESCRIPTION: Vintae launched a family of wines for the Asian market to be paired with different foods, including Bienbebido. This presented Moruba with the opportunity to create something markedly Spanish, incorporating the aesthetic of old-fashioned, authentic bars with handwritten menus—in their opinion, an unequivocally Spanish touch. A wine bottle seemed the ideal backdrop for this design. Incorporating rhyme was another way to pay homage to these old-time bars, and Moruba referenced the folk wisdom that originates in bar culture. A touch of humor complements the desired patriotic tone, as well as the actual concept of the Bienbebido packaging.

Your portfolio features a lot of handmade and organic design elements (and a lot of liquor clients), but this one is one of a few that use hand lettering. What factors made you decide to create this by hand?
JAVIER EUBA: With this project, a handmade solution was necessary to faithfully reproduce the aesthetics of bar windows. It is not a question of trends or tastes; the project needed this solution for conceptual reasons.

The design is based on traditional menu and sign lettering, correct? Can you talk about that tradition a bit? What makes the style distinctly Spanish? And how did you adapt this style to make it feel contemporary?
Painting on glass is very Spanish. I don't know the origin, but it gives the bars personality, and even if the writing isn't very clear, or the drawings are rather rough, they suggest authenticity.

Removing it from its native environment, I think, is what makes it contemporary; decontextualizing a source makes it interesting. In this case, I feel what is contemporary is the image the wine transmits—it tells a story, and more and more, this is what it is all about: telling stories in an interesting way.

This wine is targeted at the Asian market. Is that also addressed in your design choices? If so, describe.
When we create a design for the Asian market, it is important to understand the idea of the product, its origins, and its goals; you can't sell a Spanish wine in China using an Asian aesthetic. And so, for this project, we have examined the icons that clearly identify Spain in the eyes of an Asian consumer—perhaps "tapas" and "fiestas." That is Bienbebido.

Your sketches show multiple drafts of each word. Do you pick one or composite your favorite letters into a new whole?
Each draft is done in a different way: the octopus bottle has the same elements as the chicken one, but everything else is redrawn in each of them. That was part of the concept, to make each bottle genuinely different from the next. Each of them is drawn, scanned, and then silk-screened on the bottle.

Traditional Spanish menu signs were one of the
inspirations for the packaging for this wine, which
was created for the Asian market. Designing for this
demographic required the use of familiar Spanish icons.

BIENBEBIDO
- Bebe Siempre
- VINOS Y COMIDAS -
- desde siempre -

BIENBEBIDO
- desde siempre -
- 2013 -

BIENBEBIDO
- Vinos y comidas
* DESDE SIEMPRE *

HOY:

Come pollo
y
Bebe Vino

BIENBEBIDO

ABRIR POR AQUI

ABRIR (POR) AQUI

BIENBEBIDO
Bienbebido

BIENBEBIDO
* Hoy *

— Hoy —

Come PULPO
y Bebe
VINO

* HOY *

Come Cerdo
y Bebe
VINO

— Hoy —
Come VACA
y
Bebe VINO

Come
Pollo
y Bebe Vino

Bienbebido's style is derived in part from lettering found on the windows of Spanish bars. Moruba's sketches depict playful, simple illustrations and a variety of lettering that reflects its local sources.

Bienbebido's slightly roughened look was achieved naturally, using a paintbrush to render letterforms and simple illustrations.

SINISTER SWEETS
Two Arms Inc.

T wo Arms Inc. is a Brooklyn-based husband-and-wife design team specializing in custom illustration and typography, with a range of clients including the MoMA Design Store, *Time Out* New York, Nike SB, Jack Daniels, *Maxim* magazine, Jameson, and Warner Music Group.

PRODUCT: Sinister Sweets
CLIENT: Bath & Body Works
AGENCY: Chad Lavigne LLC
DESIGN FIRM: Two Arms Inc.
ART DIRECTOR: Joshua English
DESIGNERS: Michael Tabie and Karen Goheen
HAND LETTERER: Michael Tabie
MEDIUM: Brush pen
COUNTRY: United States

PROJECT DESCRIPTION: Bath & Body Works puts out a variety of holiday-themed illustrated packaging. Two Arms created the identity for Sinister Sweets, a set of three seasonal body lotions that includes Wickedly Spiced Pumpkin, Black Candied Apple, and Poison Plum.

How did you arrive at the different styles of typography for each bottle?
MICHAEL TABIE: After working through a variety of design directions, we realized the inherent layout issues we were faced with given the shape of the label. An illustrated scent name allowed us to make the type larger and more prominent.

What were the instructions from the client?
The project was art directed by Joshua English, of Chad Lavigne LLC. Our creative brief was very loose. The original concept for Sinister Sweets was a "Sinister Soiree party invite." After working on a few different holiday launches, we knew that keeping the flavor name larger was always a plus.

How do you achieve the rough textures that are a hallmark of your work?
Our textured style comes from our background as screen printers. We like to take advantage of texture and halftones as a way of working with limited color palettes.

As a printer you want to get the most out of each color, and limit the number of times you have to go through the press. This texture style has since become part of our design toolbox.

Why did you choose such loose and fun lettering rather than blackletter or movie monster type?
In this case, the hand-drawn style of the lettering fits the curves of the bottle better than a more rigid or straight typeface. In keeping with the playful nature of Halloween-inspired flavor names, we went with a more simple execution of fruit silhouettes as a holding device for the lettering.

Bold will hold. "While we always try to sketch out as many ideas and variations as possible," Tabie says, "sometimes the best answer is the first answer. This was one of those rare occasions when our client loved our first version."

173

Hand-drawn letters prior to the addition of color and gritty background texture connote the flavor of Halloween.

BUCK O'HAIREN'S LEGENDARY SUNSHINE

Device Creative Collaborative

Device Creative Collaborative is a design studio based in Winston-Salem, NC. They honor craft and get their hands dirty as often as possible.

PRODUCT: Buck O'Hairen's Legendary Sunshine
CLIENT: Sunshine Beverages LLC
DESIGN FIRM: Device Creative Collaborative
ART DIRECTORS: Shane Cranford, Ross Clodfelter, and David Jones
DESIGNERS/HAND LETTERERS: Shane Cranford and Ross Clodfelter
MEDIUM: Pencil and marker
COUNTRY: United States

PROJECT DESCRIPTION: Sunshine Beverages LLC came to Device to design a unique package that would stand out among the many other energy drinks out in the market. The design needed to reinforce the history of the legendary Sunshine recipe, created by Buck O'Hairen, a notorious nineteenth-century Appalachian moonshiner.

Why does beverage packaging lend itself readily to hand-drawn type?
SHANE CRANFORD: Buck O'Hairen's Sunshine is a pick-me-up developed from a recipe created during the 1800s. The design is meant to harken back to the hand lettering that was popular during the boom of medicine shows throughout the United States in the nineteenth century. Buck O'Hairen's Legendary Sunshine lends itself to being hand drawn as a point of differentiation. This product is not Red Bull, and is not meant for those jumping from space or doing back flips on motorcycles.

Can you talk about the neo-Victorian style that you chose for the packaging?
The Victorian style has had a huge resurgence in popularity, with many designers putting their own twists on it. For us, though, it's not just a style. The Victorian design direction conceptually tied into what this brand represented.

Where do you look for inspiration for lettering?
I find lettering inspiration everywhere. I know that is the expected answer, but it's

true. One of my favorite things to do is to rummage through old bookstores and antique shops. For online inspiration, I use some of the more popular options like Pinterest, Instagram, Dribbble, etc.

Were there many rounds of sketches created before you arrived at the final concept?
This package design came together pretty quickly. The overall layout and design details, from the sunshine monogram to the American flag barcode, were conceived right from the start. What took the most time was continually resizing the design, since choosing a final can size took some time.

Is the typography created from traced letterforms, or was it done completely freehand?
For this project, the lettering is about half freehand and half referenced from existing type. The main elements were roughly sketched out a few times then scanned and cleaned up. The remaining elements either used digital type as reference or were set entirely in digital type.

Buck O'Hairen's Legendary Sunshine is the antithesis of other energy drinks, and the design reflects this. "No extreme bevels or glow effects were used in the making of this package," says Cranford.

Even the American flag gets a whimsical touch of design on this hand lettered label.

DRINK THE SUNSHINE.COM

The BUCK O'HAIREN'S

HB

FLAVORED

NATURALLY

Legendary

SUNSHINE

REG. — TRADEMARK

Ginger Berry

FLUID OUNCES 8.4 ★ 250 MILLILITERS

XXX

≡ INSPIRED BY ≡

The Original Recipe

OF 19 TH

CENTURY MOONSHINER

BUCK O'HAIREN

SUNSHINE IS A

Delightfully Crisp,

LIGHTLY CARBONATED

PICK-ME-UP

WITH ELECTROLYTES

VITAMIN B 12

Natural Ginger

AND FLAVORS OF

BLACKBERRY.

It'll

FIX YER MORNIN'

★ RIGHT! ★

NUTRITION FACTS

SERV. SIZE 1 CAN
8.4 FL OZ (250 ML)
CALORIES 60

*PERCENT DAILY VALUES (DV) ARE BASED ON A 2,000 CALORIE DIET

	AMOUNT/SERVING	%DV*		AMOUNT/SERVING	%DV*
	TOTAL FAT 0G	0%		POTASSIUM 70MG	2%
	TRANS FAT 0G	0%		TOTAL CARBOHYDRATE 15G	5%
	SODIUM 35MG	1%		SUGARS 15G	
	CHOLESTEROL 0MG	0%		PROTEIN 0G	0%

VITAMIN B12 50% • CALCIUM 5% • MAGNESIUM 2%

NOT A SIGNIFICANT SOURCE OF VITAMINS A, C, IRON OR DIETARY FIBER. NOTE: CAFFEINE 50 MG

INGREDIENTS: FILTERED CARBONATED WATER, SUCROSE, NATURAL FLAVORS, CITRIC ACID, CALCIUM LACTATE, MAGNESIUM CITRATE, SODIUM BENZOATE AND POTASSIUM SORBATE (PRESERVE FRESHNESS), MONOPOTASSIUM PHOSPHATE, POTASSIUM CITRATE, MALIC ACID, CAFFEINE, STEVIA REB A 99% CARAMEL COLOR, SALT, CYANOCOBALAMINE.

5 03242 34343 0

"It's always interesting when asked for process sketches of a particular design," says Cranford. "It seems like most would show a linear development of a concept from start to finish. As you can see, these do not."

Ginger Berry

LIGHTLY CARBONATED 7.5 :222: MILLITERS

The BUCK O'HAIREN'S
Legendary

DRINK THE SUNSHINE.COM

LEAP ORGANICS BODY BAR
Moxie Sozo

Moxie Sozo is a world-class design and advertising agency located in Boulder, CO. The agency's accounts range from start-ups to global brands like Nike, Adidas, Nickelodeon, Mazda, Jaguar, and Campari. In addition to being one of the hottest small shops in the country, they were the first agency in the world to become zero waste, carbon neutral, and powered by renewable energy.

PRODUCT: Leap Organics Body Bar
CLIENT: Leap Organics
DESIGN FIRM: Moxie Sozo
ART DIRECTOR: Leif Steiner
DESIGNER/HAND LETTERER/
ILLUSTRATOR: Charles Bloom
MEDIUM: Pencil, acrylic, watercolor, and digital
COUNTRY: United States

PROJECT DESCRIPTION: Moxie Sozo was asked by Leap Organics, a bath and body products company, to design the packaging for three varieties of soap in Leap's line of bar soaps. The visual language varies for each package (depending upon the origins and effects of its ingredients), and inspiration was drawn from sources as varied as Cuban cigar boxes, French Deco, and penny dreadfuls (Victorian-era pulp fiction). All sides of each box are covered with slightly bizarre illustrations and quirky text, resulting in a unique trio of boxes possessing their own personality and sophisticated whimsy.

When working with the client on a project that is 100 percent handcrafted, how do you manage edits and changes, especially to the type?
CHARLES BLOOM: The first thing to do is to simply see it coming. Design, especially packaging design, needs to be flexible to accommodate various issues that may arise, such as legal requirements or when clients change their minds. I keep many of my elements separate, so they can be repositioned or scaled slightly to adapt to the creative process. Very refined sketches early in the process can save some headaches. As art progresses from sketch to final rendering, opinions change and new problems need to be solved. Inevitably, some art needs to be recreated, and it helps to mentally anticipate such edits. Of course, it also helps if you enjoy what you're doing!

How many rounds of changes occurred? Were other aesthetics explored?
How many rounds of changes? Countless! This was a labor of love (on both the client side and the design side), so the emphasis was on getting it done right, not getting it done fast. From beginning to end, the evolution of the project was quite astounding. One of the earlier incarnations featured a more graphic, digital style, depicting robots, ninjas, and dragons—a far cry from the flora and fauna that adorn the end product. The final, handcrafted aesthetic imparts a premium and natural quality that aligns with the product itself.

It appears that this is all done in-house. What are the advantages over hiring an illustrator and letterer?
By keeping the design, illustration, and lettering in-house, the concept and artwork have the ability to grow in a different way than they would have had the work been outsourced. Logistically, it's a simpler process. And because the stylistic influences vary so greatly from package to package (and sometimes within each package), the inherent cohesion that comes from execution by the same hand prevents the end result from looking too scattered or dissonant. This holistic approach yields a more refined product.

Each package was created by the same illustrator, Charles Bloom, in part to make sure every package felt cohesive within the product line.

LEMONGRASS

·ORANGE· 4oz +·LIME· 113g

LEMONGRASS
SWEET ORANGE

LEMONGRASS
· SWEET ORANGE ·

EXOTIC

Bloom says, "Each package required extensive research and exploration to match the character of the product within. The citrus profile of the Lemongrass bar conjures a tropical mood, so much inspiration was drawn from the wildlife of South America and the design aesthetic of twentieth-century Cuba."

For each package in the series, a variety of compositions and content was explored. Special attention was given to custom lettering that would reflect the unique personality of each scent.

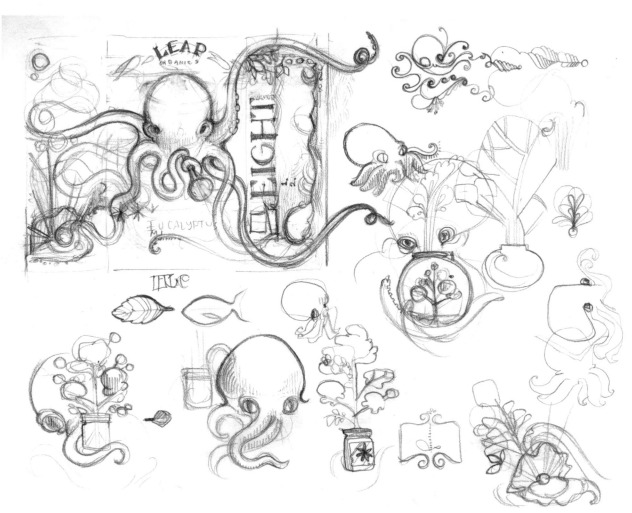

ALTERRA SALADS
Beetroot Design Group

Beetroot Design Group, based in Thessaloniki, Greece, is a multi-award-winning communication design office and think tank that provides design services and solutions to a worldwide clientele.

PRODUCT: Alterra Salads
CLIENT: Alterra S.A.
DESIGN FIRM: Beetroot Design Group
ART DIRECTOR/DESIGNER/HAND LETTERER: Alexandros Nikou
PHOTOGRAPHER: Konstantinos Pappas
MEDIUM: Ink and digital
COUNTRY: Greece

PROJECT DESCRIPTION: Alterra S.A., an innovative frozen fruit and vegetable company, commissioned Beetroot to design the packaging for their salad range. The packaging had to be neat and smart, while somehow communicating the unbroken connection between the frozen product and the natural environment in which it is grown.

What does hand-drawn type say about Alterra that typeset fonts cannot?

BEETROOT: Handwritten typography captures the human element more than any premade set of characters. In the case of Alterra Salads, our goal was to instill this particular human aspect as a subtle connection between the products and the human beings who grow, gather, package, and consume them. But we didn't want the design to be too graphic (e.g., showing farmlands, packaging plant employees, or cute little kids who "eat their greens").

You mention wanting to create a "homemade, friendly, Mama's cooking feel." The rest of the package is very crisp and bright. Was this done to create tension between modern and casual? Or was it part of the existing Alterra brand?

Setting the handwritten typography as the main communication device within a clean composition was our deliberate choice. We aimed to strengthen the visibility of our elements and to keep them clear of obstructions, or from colliding with each other. The sizing of those elements was also carefully adjusted so that the handwritten type and the sincere photograph of the product were the main visual focus. All the elements are floating freely in space and the packaging becomes casually inviting.

You frequently use custom lettering in your projects. Do your clients request it, or is it something you convince clients to use?

Usually clients are against handmade elements because, in many cases, it's much more convenient to use a visual system that can be easily reproduced rather than being dependent on someone with a "good hand."

Still, handwritten typography is not "the" way. It is a communication device with pros and cons, and when we incorporate it into a design, it's because we strongly believe that it will work to the benefit of the client's goals. Sometimes it's important to remind people that behind an image, or in this case a word (that is also an image), there is another human being. It really depends— this call is primarily ours to make—but we are always working toward the creation of a visual-producing system in which the handmade elements are incorporated organically.

Why is it important to remind a client that there's another human being behind an image?

Clients do not have to be reminded of anything. Clients need results according to their needs and specifications. It's the users, consumers, and target audience who we are trying to reach. By introducing processes and outcomes with a visible human factor, we remind users that everyone involved in the communication process, including the clients and the designers, are human beings with an aspiration to make connections.

These hand-drawn characters seek to create a human connection between the frozen product and the natural environment. The combination of casual lettering and illustrations further evokes the "Mama's home cooking" aesthetic, as if jotted down on a well-loved recipe card.

RUFA
Senyor Estudi

Senyor Estudi is a two-person company founded by Lluís Serra and Mireia Sais. Serra studied graphic design in Barcelona, and Sais studied industrial design in Girona, Catalonia. The Senyor Estudi office is located in a small village, far from Barcelona, in the heart of the Empordà, in the Costa Brava, a coastal region in northeastern Spain.

PRODUCT: Rufa
CLIENT: Rufa Cervesera
DESIGN FIRM: Senyor Estudi
DESIGNERS: Lluís Serra and Mireia Sais
HAND LETTERER: Rob Verhaart
PHOTOGRAPHER: Roger Lleixà
MEDIUM: Chalk and paint
COUNTRY: Spain

PROJECT DESCRIPTION: Rufa is a craft beer made of local and organic ingredients. This area is located on the border between Catalonia and France, in the heart of the Costa Brava (featured in Best Trips 2012 by *National Geographic*), and the land yields the finest ingredients—such as the Florence Aurora wheat, a native variety of the Aiguamolls de l'Empordà (Empordà wetlands)—to make a high-quality craft beer. Additionally, organic honey or a touch of sweet Grenache wine from the Empordà are used in the formula.

The ingredients are closely linked to the Empordà region. The bees that create the organic honey are located in the natural parks of the area. The sweet wine called Garnatxa de l'Empordà is traditional, native, and unique. The Florence Aurora wheat is harvested just before brewing. It is a very old variety of wheat that has been cultivated in the area for many centuries. The name Rufa refers to a particular cloud, which is closely linked to the Tramontana wind, a local wind from the north.

You use a considerable amount of custom type in your design. What is your process in terms of how you choose a typographic direction?
LLUÍS SERRA: Paella dishes, daily menus, and fresh beer are the main items the restaurants and bars on the Costa Brava write, in calligraphy, on their chalkboards. Therefore, we thought that it would be appropriate to take inspiration from them. This graphic style did not correspond with the usual aesthetics for traditional beers, but these are not traditional beers.

Café chalkboards are common across the world. You cite the Costa Brava region in Spain. Is there anything specific to the lettering styles of the region?
The diversity of chalkboards and restaurant menus around the world is exciting. I am convinced that trends can be found in each region, but today, everything is very chaotic and amateurish. In fact, much depends on the waiter's handwriting style. There are very interesting books and studies about the popular graphics, signs, and lettering from Barcelona, such as *Barcelona Gràfica* by América Sánchez. However, in this case we did not go that deep into its history. We just wanted to convey this diversity. Each beer is visually different but maintains the aesthetics of the chalkboard as a common element.

This is quite different stylistically from the North American chalkboard trend. Were you aware of that style? And did it affect your choices?
I would like to give you a more interesting answer, but I would be lying if I told you that the hand-drawn type is related to the essence of a main local style. It is mainly inspired by chalkboards that can be found on the coast—some of the boards directly scanned, others reproduced—but it does not mean that they represent a specific style. We were particularly interested in the diversity. We knew that the product range would grow, and that allowed us to explore different languages and different techniques. For example, for the latest beer that was released on the market, Rufa Blat (wheat Rufa), an agreement was signed with the Girona Baker's Guild to use a variety of wheat that they harvested. Therefore, we wanted the label to convey the effect of the glass used in old bakeries, on which the "deals of the day" were painted with a brush and chalk. So, we explored this technique, but in this case, we reproduced typography with calligraphic origin.

There is a lot of text on the bottles. Did you create a complete alphabet and then composite that? Or did you draw every word by hand?

OPPOSITE: *"The first three editions of Rufa are just the beginning," says Serra. "The graphic style for the labels is designed with the hopes of expanding the range of products using the same concept. The constraints are chalk and chalkboard, but the content varies."*

All the information, including legal, is drawn. We did not want the required information to seem required. So we drew it all, even the barcode. There is no computer-set text. We were interested in the texture of the chalk or the marker, so it would not have made sense to trace (vectorize) it. There are some cases in which the information varies from one year to another, such as the alcoholic content, for which we have already drawn all the numbers. This will allow us to keep changing it every year without having to lift a brush again.

Rufa solera

CERVESA ARTESANA AMB GARNATXA 33CL
DE L'EMPORDÀ 8% VOL
ALTA FERMENTACIÓ
ELABORADA A BASE
D'AIGUA, MALT D'ORDI
BLAT, FLOR DE LLÚPOL,
LLEVATS I GARNATXA
DOLÇA

0 4. GEN. 2014

ELABORADA PER RUFA CERVESERA DE L'EMPORDÀ SCP
FIGUERES. SPAIN. RSIPAC: 30.05497/ CAT. POT CONTENIR
POSIT DEGUT A LA SEGONA FERMENTACIÓ A L'AMPOLLA
LOT I CONSUMIR PREFERENTMENT ABANS DE A L'ETIQUETA

8 437012 604018

Rufa Blat

dels aiguamolls
de l'empordà

CERVESA ARTESANA
D'ALTA FERMENTACIÓ
ELABORADA A BASE DE:
AIGUA, MALT DE BLAT
FLORENCE I D'ORDI,
AURORA 33CL
FLOR DE LLÚPOL I
5%VOL LLEVATS

flequers
artesans
de les comarques gironines

03 MARÇ 2014

8 437012 604025

Rufa utilizes a variety of type styles that feel casual and unpretentious, reflecting the hand-drawn lettering found on café chalkboards throughout Europe. The packaging has a graphic and contemporary aesthetic, rather than historical.

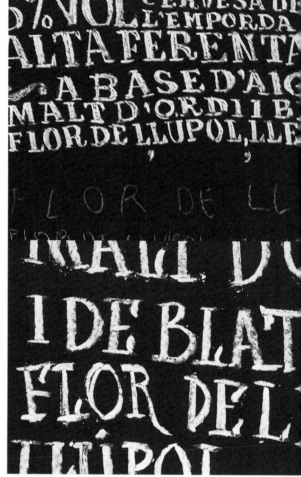

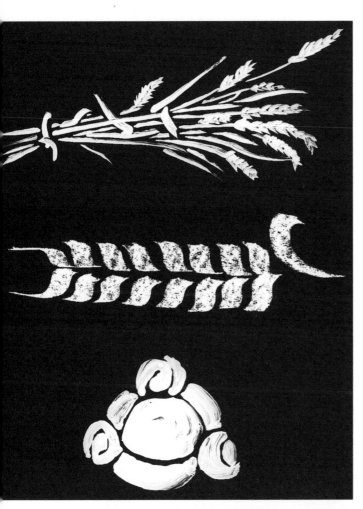

Rufa
Blat
Rufa

RVESA ARTESAN
LTA FERMENTACI
BORADA A BASE D
UA, MALT DE BLA
ENS AURORA I OR
DE LLÚPOL I LLEVA

CERVESA ARTESANA DE
GARNATXA DE L'EMPORDÀ
CERVESA ARTESANA DE
G GARNATXA EMPORDANESA

CERVESA
ARTESAN
GARNATX

33CL 5% VOL
3
33CL 5% VOL
3 33CL 3 5
3 3 3 3
CL % VOL

Arti

Sanal

Well-manicured type treatments are a joy to behold. The attention to fine detail and craftsmanship makes them the stuff of great envy (a trait that we all politely try to hide, of course). The artisanal work shown here is design done on the head of a pin, with exacting precision. It makes the transition from pure hand to digital seamlessly, feeling neither inauthentic nor overwrought.

PERFETTO PENCILS
Louise Fili Ltd

Louise Fili Ltd is a New York City–based design studio specializing in everything to do with food, typography, and all things Italian.

PRODUCT: Perfetto Pencils
CLIENT: Princeton Architectural Press
DESIGN FIRM: Louise Fili Ltd
ART DIRECTOR: Louise Fili
DESIGNERS: Louise Fili and Spencer Charles
HAND LETTERER: Spencer Charles
MEDIUM: Pencil and digital
COUNTRY: United States

PROJECT DESCRIPTION: Louise Fili loves her collection of vintage Italian pencil boxes, many still filled with their original pencils. Her most preferred are the two-color, double-sided pencils, commonly available in red and blue, for teachers to correct homework with: "*Errore lieve, segno rosso; errore grave, segno blu*" (red for a minor infringement, blue for a serious offense). When Princeton Architectural Press invited Fili to come up with a line of gift products, the two-tone pencils seemed perfect—thus the name, Perfetto. Steering clear of blue, her least favorite color, she opted for her signature red and black. The design is based on pasticceria papers (used to wrap pastries) and hand-lettered upright scripts, both popular during the 1930s.

What is your work process? Do all of your projects begin with a tightly hand-drawn sketch?
LOUISE FILI: Yes. I love the sketch stage. This project was done on a very tight deadline—I had to come up with a name and then design the box, all in about a week. Fortunately, it all came together quickly and I managed to capture what I was looking for in the first sketch, which I then discussed with my senior designer, along with specific references for the script lettering, as well as the additional text.

How do you balance the old and the new, so that your work never seems to feel like pastiche?
My design language is often informed by vintage work, but my goal is to make it my own in some form. The Perfetto package draws from a number of sources—I think of the final product as my interpretation of these references.

Is this the first offering in what will be a Louise Fili line of merchandise? What's next?
A set of colored pencils called Tutti Frutti!

In what way did the Doris pencil box provide inspiration?
It wasn't any box in particular (I have many). Rather, it was the general idea of a pencil box as an object of desire. The concept of two-colored, double-sided pencils made the project particularly engaging.

Do you now use your own pencils?
Always! I never realized how useful they could be. I like to write lists in black, then circle or check off important items in red. People have responded very strongly to Perfetto—in this computer-driven world, something as prosaic as a pencil is extremely satisfying.

It was very exciting to see these for sale at the bookshop at the Palazzo delle Esposizioni in Rome!

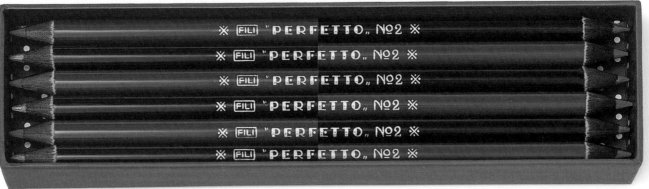

"Quality" and "elegance" are the perfect descriptors of both the Perfetto box and its contents.

"Perfetto"

"Perfetto"

"Perfetto"

Fili's sketch and the final result are remarkably similar. The Perfetto
type evolved from a monoline to an outline to an inline.

QUALiTY ELEGANCE
QUALiTY ELEGANCE

HABIBIS
Anagrama

Anagrama is an international branding, architecture, and software development firm with offices in Monterrey, Mexico, and Mexico City. Their clients include companies from varied industries in countries all around the world. Besides their history and experience with brand development, they are also experts in the design and development of objects, spaces, software, and multimedia projects. Since their creation, Anagrama has decided to break the traditional creative agency scheme, integrating multidisciplinary teams of creative and business experts.

PRODUCT: Habibis
CLIENT: Fuad Zetuna
DESIGN FIRM: Anagrama
ART DIRECTOR/DESIGNER/HAND
LETTERER: Anagrama
PHOTOGRAPHER: Caroga Foto
MEDIUM: Hand drawn and digital
COUNTRY: Mexico

PROJECT DESCRIPTION: Habibis is an Arabic-Mexican fusion taquería located in San Pedro Garza García, Mexico, a city enriched by the culinary treats of its third-generation Arab immigrants. Previously a humble taco stand, Habibis approached Anagrama with the task of creating a brand that communicated the food's mixed background and exceptional quality without losing its street-friendly and casual demeanor.

Their proposal was a brand that adapts stylized Arabic calligraphy to a typical Mexican street setting, complete with neon colors and inexpensive materials, like kraft paper bags.

Deep research and careful understanding of the Arabic alphabet was needed to design the various words and signage in both Arabic and Latin, using calligraphic pens and special brushes. The custom type is accompanied by Gotham, a gentle and neutral typeface that allows the bespoke logotypes to stand out above everything else. The pattern is based on traditional *keffiyeh* (a Middle Eastern headdress fashioned from a square scarf) and gorgeously intricate mosaic patterns.

Evoking an Arabic calligraphic style to render Latin letterforms is a delicate task; it could very easily be perceived as satire or as a condescending cartoon. How did you manage to avoid that?
ANAGRAMA: We avoided cliché by observing the strokes and gestures of Arabic calligraphy very carefully. We managed to create the type by paying close attention to those details and having respect and good taste for the task at hand.

Did you have a familiarity with Arabic calligraphy, or did you do it strictly by practicing the typical strokes and forms?
We'd never worked with Arabic type before; this was completely new for us. Since then we have had several more projects that use Arabic letterforms.

How has the Arabic community reacted?
Positively. The project was successful on Behance and on many online blogs.

Mexico has a wonderful tradition of hand-lettered signs. Do you feel inspired or connected to this tradition?
Sure. Years ago we worked on a project, la Fábrica del Taco, a Mexican restaurant in Buenos Aires, Argentina, that revolves around the hand-lettered concept. It was a very gratifying project mainly because we consider our gastronomic tradition as one of the greatest Mexican cultural legacies, and it has become the calling card for our identity in other countries.

"Simulating a style of writing that was unfamiliar was a learning experience that was quite entertaining," says Anagrama. "We used a different approach that started out as being very manual. We had to explore a lot with different materials, such as black ink and paper."

(GRACIAS)   (HABIBIS)

(BAÑOS)   (HABIBIS) T. 1768.2259

شكراً

(GRACIAS)

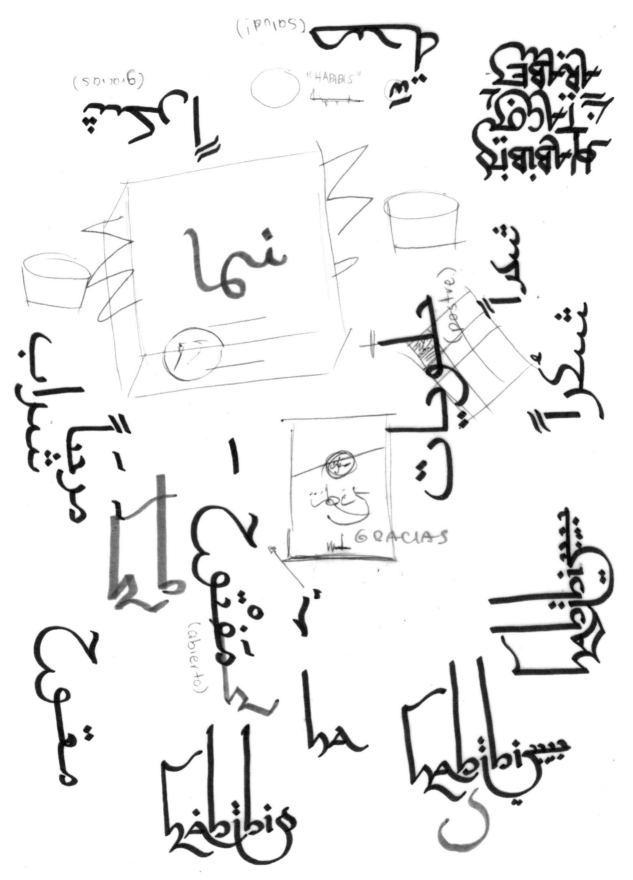

SALSA FRIJOLES

Anagrama designers watched a lot of YouTube videos about Arabic calligraphy—how to hold the brush, what direction to paint in, the position of the hands—in order to give it the proper feel. "We had to learn a few Arabic words to understand the language and how it behaves graphically," says Anagrama.

TACOS ARABES FRIJOLES JOCOQUE

حلويات

شَراب

GOLD TOP CIDER
Simon Walker

Simon Walker is an Austin-based one-man studio specializing in designs that feature custom typography.

PRODUCT: Gold Top Cider
CLIENT: Austin Eastciders
DESIGNER/HAND LETTERER: Simon Walker
MEDIUM: Pencil and digital
COUNTRY: United States

PROJECT DESCRIPTION: Walker was asked to create a label design for a new cider product, Gold Top Cider, which would eventually become the flagship for a whole line of ciders. The client was particularly interested in finding a way to recognize the existing, modern design aesthetic prevalent in Texas-based products (particularly those based in Austin), but with a heavier push toward something that felt truly authentic and from another time in US history.

You have a particular talent for capturing the flavor of vintage without seeming beholden to it. Describe your influences and how they make their way into your design.
SIMON WALKER: I'm incredibly influenced by old movies, and I think my love of vintage cinema coincided with the recent revival of vintage type that I was seeing online a few years ago, especially on Dribbble. So many great talents on there; I couldn't help but aspire to what others were doing—and have done in decades past.

Is there a lot of handwork in the development process? Or mostly digital?
Mostly digital. I find that my brain has developed a sort of muscle-memory adjustment these days in favor of vectoring. The process of clicking out a design in Illustrator and then refining the lines point-by-point feels just like the sketch-and-erase process you might find when using a pencil.

OPPOSITE: *When creating custom word marks, Walker often goes far beyond designing the necessary letters, sometimes crafting entire alphabets.*

Your custom letters often have the thorough logic of typefaces. Do you develop whole alphabets for each project or just solve things on a per-word basis?
A little of both, actually, which is to say I start out developing a letter style using only the letters required by the project, but what often ends up happening is that I develop more of those letters—either for another part of the project or a new project entirely—until I have about 75 percent of an alphabet (lower or uppercase, rarely both) complete.

You have a font based on your Dirty Jean Co. logo. Is that finished? Any others planned? How does font design come full circle and influence your lettering? What did you learn?
That font has been in the works for years now, mostly because I just haven't had the time to devote to finishing it. That's about to change though, and I think I'm close to finally releasing it into the world.

I actually do have plans to develop a number of other type styles from past projects into fonts as soon as I can get around to them. Designing letterforms has really taught me a ton about letters, in ways I simply wouldn't have grasped just by looking at type. In trying to recreate letters, I've discovered nuances and subtleties about spatial relationships between letters in a magically organic way. It becomes a kind of unwritten rulebook in your mind—and once you've learned those rules, that's when you can really have fun and start breaking them.

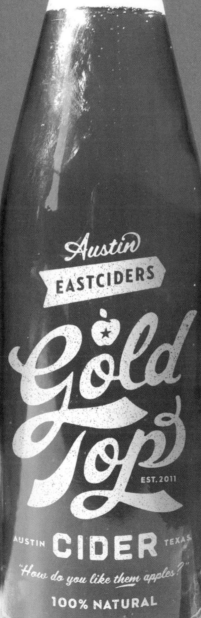

Gold Top Cider wanted something that felt true to its Austin roots and was rich in American history. Slight variations are shown on a logo that satisfies the client's needs, infusing a modern Austin aesthetic with warm Americana.

Walker equates clicking out a vector design then refining its points to the traditional sketch-and-erase process of using a pencil. While he still first draws in pencil, all refinements are created digitally.

LA SOCIÉTÉ PARISIENNE DE SAVONS CRÈME MAINS TRAITANTE

Daniel Pelavin

D aniel Pelavin is a New York City–based illustrator and typographic designer who works on a wide and varied range of projects, specializing in integrating illustration and lettering into effective and cogent communication.

PRODUCT: La Société Parisienne de Savons Crème mains traitante
CLIENT: Laboratoire HT26
ART DIRECTORS: Stéphane Ouaknine and Daniel Pelavin
DESIGNER/HAND LETTERER: Daniel Pelavin
MEDIUM: Pencil and digital
COUNTRY: United States

PRODUCT DESCRIPTION: The La Société Parisienne de Savons brand comprises twenty different themes applied across a variety of products. However, since manufacturing requirements made producing individual hand creams impractical, Pelavin designed a single hand cream tube, label, and packaging to represent the entire line. An important part of the project, separate from the graphics, was choosing a simple metal tube that would crease and wrinkle with use, as opposed to the sterile looking laminated plastic tubes currently in favor.

In rebuilding this vintage brand, how much original material did you have to work from? And how much did you need to create from scratch?
DANIEL PELAVIN: There was no original material for this product except for the brand logo, which had been designed earlier from scratch. Much vintage material was researched and explored to get a feeling for the packaging, but the ultimate design came directly from the rough concept sketches.

You seem to be a master of myriad historical lettering styles. How much do you rely on reference, and how much do you just have in your head?
You can only rely on typographic reference up to a certain point. Most samples do not contain the full complement of characters, so you have to create new ones based upon the feeling of the characters you have on hand. I prefer to create original lettering that may have the "flavor" of a time period rather than copy or appropriate existing forms.

What were the particular challenges of working in this style?
As above, creating a homogeneous suite of letterforms from scarce reference is a challenge but, for me, an engaging and welcome one.

You have been lettering for a long time now. What's your perspective on the current revival? What do you think of the more "casual" techniques some people favor?
I've been observing this "revival" for over forty years. The "casual" (what a nice euphemism for "bad") lettering is rarely done by competent practitioners and mostly represents an ill-founded backlash against the art and craft of lettering, as well as the sterility of digital work produced by amateurs.

When doing historical style lettering, how do you preserve accuracy to the original? Do you alter designs to make them more functional for the modern market? How do modern materials and tools make it easier (or harder) to create an authentic look?
I'm not interested in accuracy or authenticity—I'm trying to solve contemporary problems using lettering that references a certain time period or genre with the goal of complementing and reinforcing the literal context. Contemporary tools are both incredibly useful and indispensible considering the time frame, pricing structure, and expectations engendered by their availability.

From an initial selection of soaps, face powders, and lip balms, the offerings at La Société Parisienne de Savons have expanded into a full line of fragranced products including candles, bath salts, scent diffusers, and hotel amenity packages.

Pelavin says, "My sketchbooks are work sketchbooks rather than display sketchbooks. They have to be that way so I can be free to not think while sketching. I just let the thoughts flow."

June 7, 2012   La Société Parisienne de Savons

crème traitante pour les mains & les ongles.
Treatment Cream for hands & nails.

JULAQUAVIT
Martin Schmetzer

Martin Schmetzer is a freelance letterer and illustrator based in Stockholm. His passion for letters started with graffiti at a young age and paved the way for the style that defines him today. For every new commission, he does a new type of letter, always handmade with signature Schmetzer curves.

PRODUCT: Julaquavit
CLIENT: Gotlands Bryggeri
AGENCY: Communicatering
CREATIVE DIRECTOR: Jens Frithiofsson
DESIGNER/HAND LETTERER/ ILLUSTRATOR: Martin Schmetzer
MEDIUM: Pencil, paper, and digitial
COUNTRY: Sweden

PROJECT DESCRIPTION: Communicatering contacted Schmetzer to design a new label for Gotlands Bryggeri's Julaquavit, a Christmas aquavit only available during the holiday season.

Your sketches are very tight. Can you talk about your process?
MARTIN SCHMETZER: I always start with drawing pen on paper when I design— rough sketching first to explore different compositions and solutions for the client to consider. Once a favorite is selected, I draw more finely, tightening the lines and honing the details before I redraw it in Illustrator.

The combination of woody slab serif and (vaguely blackletter-influenced) calligraphy is interesting. Talk about that choice (earlier sketches have lettering with more similarities to the script).
I have worked with Communicatering before, so they had quite a clear image in mind for the typography, referring to previous designs I had done. They specified that I should avoid Gothic letters with too much fracture and instead focus on a type design with more motion and curves. This ultimately led to a handcrafted, authentic style with a modern touch.

Was your design inspired by more traditional aquavit packaging?
No, not really; it was more inspired by old Christmas beer labels and traditional Swedish "Julmust" designs, with a blend of old and modern aesthetics and color choices not immediately related to Christmas. Instead of using a large Coca-Cola–red Santa Claus, they wanted more of a "Jenny Nyström"–style elf/Santa.

The Julaquavit lettering is quite consistent in the sketches, but the logo seems to vary each year. Was there a lot of development before these sketches, or does it have a previous source?
I had freedom in designing this label. The arch-shaped label and border were already set, though, since this is a shape Gotlands Bryggeri wants to use consistently now throughout their beer and spirit range.

What are some of your historical influences?
I came in contact with hand lettering through graffiti, which I believe shows in most of my work. I enjoy bombastic designs and don't agree with the saying that "less is more." Also vintage Victorian hand lettering is a big inspiration to me, with the high level of detail and diligence they had back then.

Schmetzer's precision shines on a project that turned out to be a labor of love. "It was such a fun commission," he says, "that I never even took all the extra hours I spent into consideration."

WISBY
Julaquavit

50cl
2018
42% vol

222

For this project, Schmetzer drew his rough sketches slightly more tightly than he usually would to "do my vision and concept more justice."

223

"The sketching process is most valuable to me," Schmetzer says. "The computer is a fantastic and necessary finalizing tool, but it also limits my shapes and composition if I do not start first by hand, where I can go whole hog. After this stage, I have a quite clear vision of how I want the end product to turn out. But smaller details are often changed or added along the digital process. There is something abstract about letters that attracts me. You can make each letter of the alphabet look different in unlimited ways yet still be legible. I enjoy the interaction hand-drawn typography can have to the meaning of the word and how the letters next to each other play together."

STENBERG & BLOM'S ZINFANDEL
Holidesign

H

olidesign was founded in 2006 by Boris Iochev, after he relocated from Seattle to Oslo. It is dedicated to branding—from graphic design to interior architecture—dynamically detailed, nimbly executed, and holistically designed, with a global perspective by an international team of talent. In 2013 Iochev, together with Carl Fredrik Angell, cofounded Signfidelity, a hand lettering and storefront sign painting company in Norway and Denmark.

PRODUCT: Stenberg & Blom's Zinfandel
CLIENT: Stenberg & Blom AS
DESIGN FIRM: Grid Design AS and Holidesign
ART DIRECTOR: Martin Nordseth
DESIGNERS: Martin Nordseth and Boris Iochev
HAND LETTERERS: Carl Fredrik Angell and Boris Iochev of Signfidelity
PHOTOGRAPHER: Studio Dreyer Hensley
MEDIUM: Pencil on watercolor paper
COUNTRY: Norway

PROJECT DESCRIPTION: Stenberg & Blom AS had a very clear vision for their product—unambiguously, traditionally Italian. They introduced a new Zinfandel brand, using a vintage yet contemporary style of design. With so many offerings at Vinmonopolet (Norway's state-regulated liquor store chain), Holidesign needed a way to clearly differentiate the new bottle, giving it maximum shelf presence.

What were some of your design influences in creating this packaging? What is the ornamentation based on?
BORIS IOCHEV: Remaining true to this product's origins, we were inspired by the authentic, Italian hand-painted ceramics of Deruta, which are highly collectible and prized throughout the world.

There is virtually no difference between the sketch and final. Is such accuracy and detail typical of your work?
There were some small variations of details that were not documented properly. However, hand lettering and/or painting of storefront signs is a laborious process; therefore, we strive to define the task and outcome as closely as possible. Like everything done by hand, it takes more time than pressing a command key on your keyboard.

What are the challenges of designing liquor packaging?
Finding that perfect balance between expectations and surprise. The final expression should be somewhat familiar to what a consumer may have as a most common association with that product, yet it should possess a certain amount of surprise in order to stand out on the shelf.

Can you talk a little bit about the relationship between design and music, since you're a musician of note?
Both music and design follow identical processes—research, flirting with different ideas, falling in love with one of them and finessing it further, balancing strategy with the artist's own intuition, and the transfer of excitement to the audience.

OPPOSITE: *"The wine label we crafted with so much passion has been a huge success," says Iochev. "During the first month of the launch, over five thousand bottles effortlessly found their way to different wine aficionados all around Norway. It's like selling records."*

ANNATA
2012

VITICOLTURA DI

MONT

The Italian

Zinfand

Letterforms and ornamentation were carefully drawn in pencil on watercolor paper, with a ruler used to precisely render straight lines.

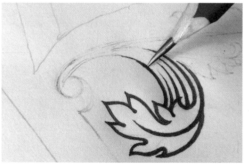

MONTICO

The Italian

Zinfandel

di Puglia

230

BUMMER & LAZARUS DRY GIN
Dave Stevenson

Dave Stevenson is a freelance illustrator who works from his home studio in Dixon, CA. He has produced illustrated maps for *National Geographic* magazine, *Outside*, and the Walt Disney Family Museum, as well as hand lettering for Sierra Nevada Brewing Company, Sonoma Wine Company, Sutherland Distilling Company, and Valley of the Moon Winery. Stevenson's work has been included in *How*, *Art Direction*, *Communication Arts*, *Print*, *American Illustration*, and the *Graphis* annual, and he has received silver medals from the Art Directors Club of New York and the San Francisco Society of Illustrators.

PRODUCT: Bummer & Lazarus Dry Gin

CLIENT: Raff Distillerie

DESIGN FIRM: Auston Design Group

CREATIVE DIRECTOR/DESIGNER: Anthony Auston

HAND LETTERER: Dave Stevenson

ILLUSTRATOR: Mike Gray

MEDIUM: Ink on vellum

COUNTRY: United States

PROJECT DESCRIPTION: Raff Distillerie is located in the middle of San Francisco Bay, on San Francisco's Treasure Island. The owner and master distiller, Carter Raff, came to the team with a very interesting brand concept for Raff Distillerie's line of products. He wanted the labels to represent the quirky and interesting side of San Francisco's Barbary Coast during the mid-to-late 1800s.

One of the great stories from this era was of two stray dogs, Bummer and Lazarus, that were beloved celebrities in the community. They were known to kill hundreds of rats a day, in a town that apparently had no shortage of them.

The approach to the design of this label was to present the stories of Bummer and Lazarus while evoking the look and feel of the era. This needed to be done in an upscale and eye-catching way worthy of a $35.00 bottle of gin. The labels are designed to look as though they were printed with the technology of the day; in this case, engraving. The majority of the type, lettering, and decorative embellishments are hand done specifically to replicate the look of a hand-engraved steel plate etching.

What is your typographic background? Are you self-taught or trained in Victorian lettering?
DAVE STEVENSON: I did study some hand lettering in college, but at that time, the hand lettering that was taught involved working with chisel point charcoal for the express purpose of doing comprehensive mock-ups for advertising layouts. Of course, those days are long gone. After working as a packaging designer in San Francisco in the late 1980s, I chose to freelance, concentrating on illustration for advertising agencies and graphic design firms.

Years later, I produced a series of map illustrations of the Panama Canal for Kit Hinrichs at Pentagram. I used those maps to market myself to publishers and magazines and created maps for *National Geographic* magazine, *Outside*, and others. After ten years, I could see the decline of the magazine industry, and those still around were increasingly relying on computer-generated map art. I often included hand lettering when painting the maps, so when I decided to return to packaging, I tailored my portfolio toward hand lettering. I saw a need for proprietary custom lettering that had a hand-drawn quality to it. I have been fortunate to find designers looking for just that.

Bummer and Lazarus were two legendary stray dogs: champion ratters in San Francisco's red light district of the mid-to-late 1800s, "The Barbary Coast."

What was the decision making process like? All of the sketches are so beautiful that I would imagine that it was next to impossible to decide.

All artwork is hand inked or painted. I want to thank you for your compliment on my sketches, but I have little or no control over the selection process. I draw based on the direction of the designer and usually include a couple of sketches with a direction that I like, and then it is in the hands of the designer and his client. That is why it is so enjoyable to work with people whom you respect and trust to do great things with your contribution to their project.

To what do you attribute the resurgence of hand lettering?

I'll take an educated guess here, but I attribute the resurgence to two factors. First, in an overexposed, slick computer-generated packaging world, where every brand manager wants their product to stand out on the overcrowded grocery shelf, some see the value of a handcrafted package. Second, there is a lively business of small microbreweries, small wineries, and microdistilleries. They, rightly so, want designers to create packaging that reflects their hands-on approach in the label as well as in the supporting marketing materials.

Extensive explorations in period style helped
define the tone of the final design.

Even the smallest details were carefully considered, with
Stevenson drawing intricate variations on information such
as the alcohol content of the gin.

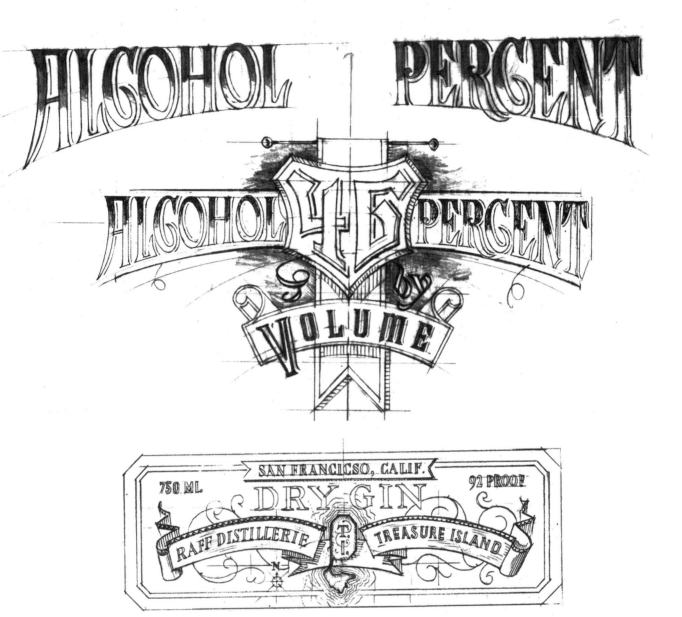

ALCOHOL PERCENT

ALCOHOL 45 PERCENT by VOLUME

SAN FRANCICSO, CALIF.

750 ML. DRY GIN 92 PROOF

RAFF DISTILLERIE TREASURE ISLAND

SAN FRANCICSO CALIF

750 ML DRY GIN 92 PROOF

BOTANICALS = CORN, JUNIPER, RYE ?

EVIL SPIRITS VODKA
Saint Bernadine Mission Communications Inc.

Saint Bernadine Mission Communications Inc. is a Vancouver-based agency specializing in branding, advertising, graphic design, and digital design.

PRODUCT: Evil Spirits Vodka
CLIENT: Evil Spirits Distillery
DESIGN FIRM: Saint Bernadine
Mission Communications Inc.
CREATIVE DIRECTORS: Andrew
Samuel and David Walker
ART DIRECTOR/HAND LETTERER/
ILLUSTRATOR: Jennifer Hicks
MEDIUM: Ink and digital
COUNTRY: Canada

PROJECT DESCRIPTION: Evil Spirits Distillery is a new premium spirits line that is painstakingly crafted to be sinfully enjoyed. Saint Bernadine was tasked with package design and a media kit for Evil Spirits' launch.

Evil Spirits Vodka is the product of a distilling craft so refined it must have been exchanged for diabolical favors. Saint Bernadine extended this brand promise through every detail of the design language. Custom word mark? Evil. Spirit renderings in matte varnish? Evil. Recipe card? Evil. Ouija board with commemorative cast pewter planchette? Evil.

Some of your sketches just say "Evil Vodka." Were you party to that name change?
ANDREW SAMUEL: One of the early explorations was inspired by the Prohibition era, and the belief that alcohol itself was evil. The distillery name would have remained Evil Spirits Distillery—it, and their products, would have been Evil Vodka, Evil Gin, etc., under the umbrella distillery name.

This project was hand lettered, but in a much different style than R&B Brewing. How does Saint Bernadine decide on an appropriate style?
The words "Evil Spirits" were created much in the same way that a word mark or logo is customized to be original and fit the brand's aesthetic. This word mark does not have the same hand-lettering feel that the R&B beers have—R&B's brand is all about having a handcrafted look. Style is largely dictated by the brief—Evil Spirits is a painstakingly crafted premium vodka, so the result needed to feel cared for and premium…and, evil.

How much do you rely on compositing with fonts before starting on custom lettering?
Not much. The early sketch word marks are sometimes inspired by other typefaces but are often imagined and custom to each bottle sketch. At a more refined stage, the resulting Evil Spirits word mark was inspired by a few different typefaces and created by tracing over existing letterforms, adding new elements and characters from other typefaces, skewing it, and tracing it again and again, trying new things until the desired look is achieved.

Were you able to show rough pencil sketches to the client or were the ideas more fully formed for presentation purposes?
Yes, things were kept rough and in sketch form until the approved Ouija concept. From there, it was a combination of a sketched label applied to a photograph of the bottle in Photoshop, with multiple color and word mark combinations done digitally after that. Once a word mark and loose color palette had been decided on, the actual digital artwork began.

Inspiration for Evil Spirits vodka comes from Ouija boards, and was born from the idea that alcohol was considered evil in the Prohibition era.

Obvs evil

"We work on mood boards as a preliminary to the design stage proper, but after we have a rough conceptual and strategic direction" Samuel says. "Mood boards are a great way to start thinking and talking about design in a concrete way without having to do much work, and as such we treat them as a pretty fluid tool, allowing us to refine concepts and design direction very quickly."

alcohol is evil / prohibtion propaganda

EVIL

E EVIL

EVIL
SPIRITS

EVIL
SPIRITS

EVIL
SPIRITS

EVIL
SPIRITS

EVIL
SPIRITS

EVIL
SPIRITS

EVIL
SPIRITS

EVIL
SPIRITS

EVIL
SPIRITS

EVIL
SPIRITS

EVIL
SPIRITS

EvilSp

EVIL
Spirits

EVIL
SPIRITS

The final Evil Spirits word mark was inspired by a few different typefaces
and was created by tracing and retracing existing letterforms, adding
customized elements.

"Once the strategic direction has been set, we start sketching" Samuel says. "Working small and in pencil allows us to focus on elements like type and layout, bottle shape, silhouette, and design themes without getting bogged down with detail, color, etc. Again, the idea here is to work fast and generate a lot of ideas quickly. Once we have sketches, we cut them up and organize them thematically, and cull the ones that work less well. The best sketches then become the basis for further refinements, and so on. We work on multiple rounds of sketches, often four or five rounds, before finally moving to computer renderings."

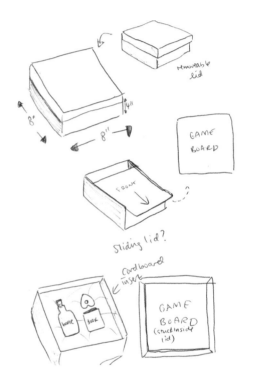

frosted band
within filled

frosted
letter

black matte
bottle w/
shiny black
over

black hard
w/ silver lines

black band
w/ silver
+ clear

black
+ silver
silkscreen

EVIL
SPIRITS

245

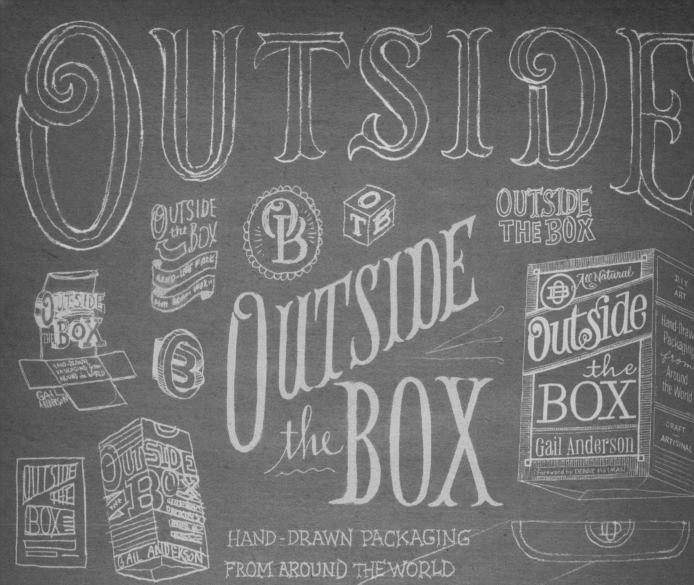

OUTSIDE THE BOX

Outside the Box

the BOX

HAND-DRAWN PACKAGING
FROM AROUND THE WORLD

GAIL ANDERSON

Foreword by Debbie Milman

INSIDE *OUTSIDE THE BOX*

Gail Anderson & Joe Newton

After spending months soliciting sketches for this book from generous designers, letterers, and studios, it's only fair to show a little of our behind-the-scenes design process, too. We always like to put pencil to paper before moving over to the Mac, and have copious amounts of drawings created mostly on recycled letter-sized printer paper. Many sketches are tightly rendered, and some are better than others, of course. Actually, a few are downright corny, but since we're amongst friends, we'll share them with you, anyway.

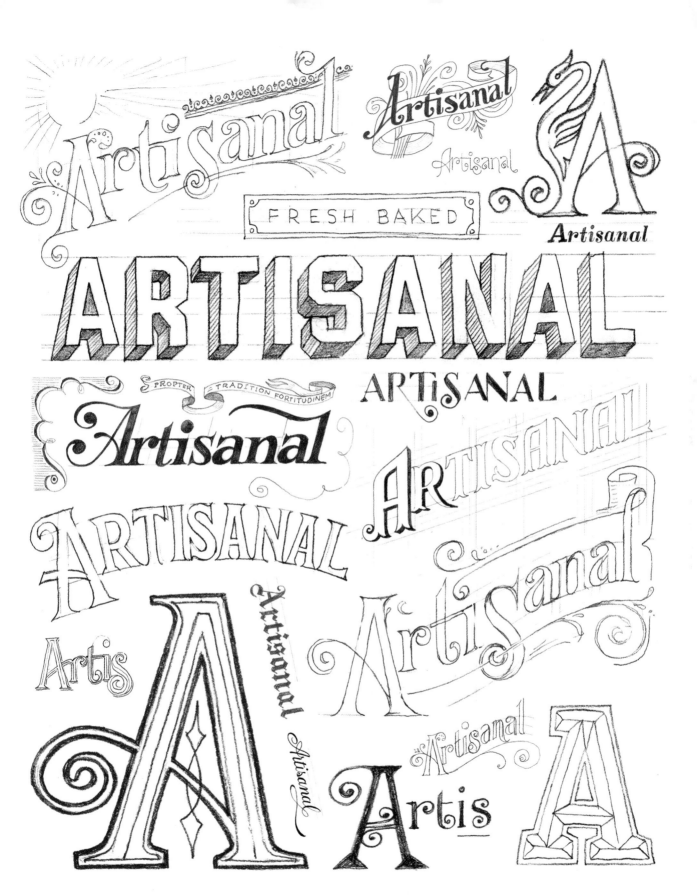

RESOURCES

Sean Freeman
http://thereis.co.uk

Jessica Hische
http://jessicahische.is

Lauren Hom
http://homsweethom.com

Seb Lester
http://www.seblester.com

Like Minded Studio
http://www.likemindedstudio.com

Luke Lucas
http://www.lukelucas.com

Erik Marinovich
http://www.erikmarinovich.com

Gabriel Martínez Meave
http://www.meave.org

Grady McFerrin
http://www.gmillustration.com

Mucca Design
http://mucca.com

John Passafiume
http://johnpassafiume.com

Mike Perry
http://www.mikeperrystudio.com

Chad Roberts
http://www.chadrobertsdesign.com

Jeff Rogers
http://www.howdyjeff.com

Nancy Rouemy
http://www.welivetype.com

Alex Trochut
http://www.alextrochut.com

BIBLIOGRAPHY

Clouse, Doug, and Angela Voulangas. *The Handy Book of Artistic Printing: A Collection of Letterpress Examples*. New York: Princeton Architectural Press, 2009.

Fili, Louise. *Elegantissima: The Design and Typography of Louise Fili*. New York: Princeton Architectural Press, 2012.

Gibbs, Andrew. *Box Bottle Bag: The World's Best Packaging Design from thedieline.com*. Cincinatti: How Books, 2010.

Heller, Steven, and Seymour Chwast. *Graphic Style: From Victorian to New Century*. New York: Harry N. Abrams, 2011.

Heller, Steven, and Mirko Ilić. *Handwritten: Expressive Lettering in the Digital Age*. London: Thames & Hudson, 2006.

Heller, Steven, and Gail Anderson. *New Vintage Type: Classic Fonts for the Digital Age*. London: Thames & Hudson, 2007.

Heller, Steven, and Lita Talarico. *Typography Sketchbooks*. New York: Princeton Architectural Press, 2011.

Levine, Faythe, and Sam Macon. *Sign Painters*. New York: Princeton Architectural Press, 2013.

Perry, Michael. *Hand Job: A Catalog of Type*. New York: Princeton Architectural Press, 2007.

Type Directors Club. *Typography Thirty Five: The Annual of the Type Directors Club*. New York: Harper Design, 2014.

GAIL ANDERSON is a New York City–based designer, writer, and educator. She is a partner at Anderson Newton Design (AND) and on the faculty at the School of Visual Arts. Anderson serves on the Citizens' Stamp Advisory Committee for the United States Postal Service, and on the board of the Type Directors Club. She is the recipient of the 2008 Lifetime Achievement Award from the AIGA.

http://www.gailycurl.com

PORTRAIT BY PAUL DAVIS

COLOPHON

This book was typeset in:
Filosofia, from Emigre
http://www.emigre.com
and
Adelle, from TypeTogether
http://www.type-together.com

T H A N K S

My special thanks is extended to Sara Bader from Princeton Architectural Press, whose idea all of this was in the first place, and to the fabulous Louise Fili, who put us together. I am also indebted to Tom Cho from PAP, whose helpful edits, guidance, and good humor made working nights and weekends a little more fun. And high praise to PAP's design director, Paul Wagner, whose keen eye to detail is much appreciated.

Thanks goes to Joe Newton, my workmate and friend, whose assistance throughout this project proved to be invaluable, as always. Joe put up with a lot of hand wringing and whining—even more than usual—and I am forever in his debt for being a such good egg and good listener.

Thank you to Debbie Millman for providing the foreword to this book, including those swell little spot illustrations. And I am extremely grateful to Steve Heller, not just for providing a quote, but for being my writing mentor. And speaking of quotes, Andrew Gibbs, Parham Santana, and the lovely Roberto de Vicq de Cumptich get special props for the wise words they wrote for the introduction.

Betsy Mei Chun Lin, Maria Sofie Rose, and Maxwell Beucler all interned on this project at various points, and their hard work and patience is greatly appreciated. A huge debt of gratitude is also extended to my old friend Gary Montalvo, who came in at the end, as he always does, to help wrap up the layouts.

Of course, this book would be a wad of blank pages if it were not for the designers, letterers, and studios who responded to my emails, sent their amazing work, and took the time to answer my questions. My deep appreciation is sent (via email) to each and every one of you, and know that you're in good company.

Finally, thank you Gerry Anderson Arango, Mike Anderson, Randi Wirth, and Laura Worthington. You know what you did, and I thank you for doing it.

GAIL ANDERSON

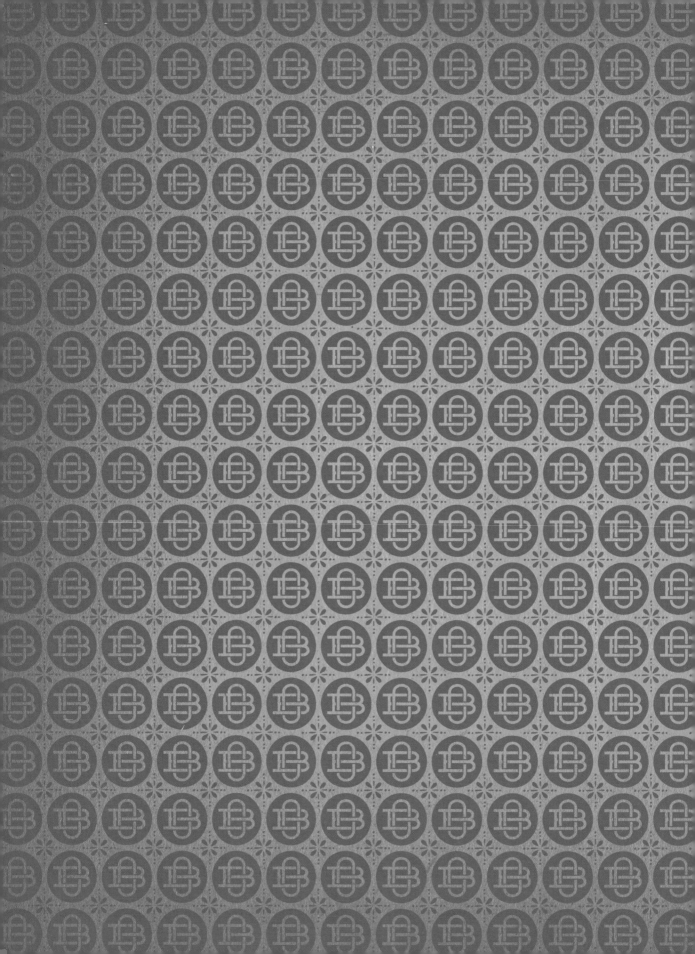